Technical English 3

Course Book

David Bonamy

Contents

1 Systems

1 Rescue

Start here 1 Work in pairs. Answer the questions about the safety devices in this illustration of an air-sea rescue.

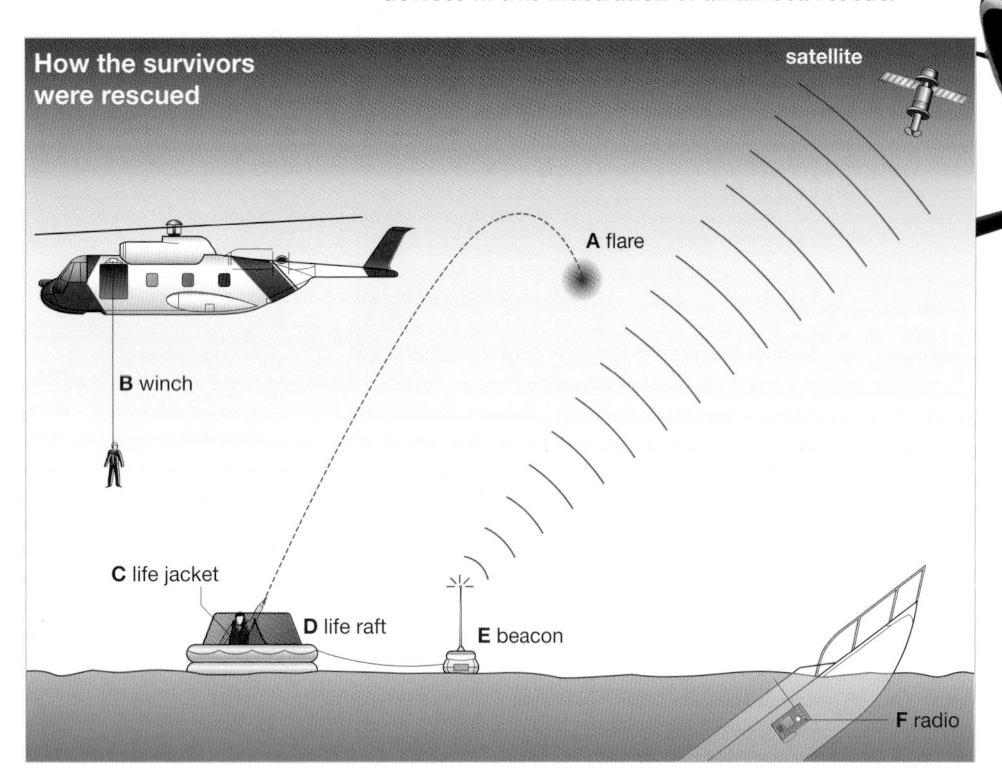

Which ones:

1 stop you from sinking?
2 tell the rescuers where you are?
3 rescue you from the water?

Listening 2 02 Listen to this news report and put the six safety devices from 1 in the order the reporter mentions them.

1 ___ 2 ___ 3 ___ 4 ___ 5 ___ 6 ___

3 Put these statements in the order the events actually happened. Then listen again to check your answers.

A ______ The helicopter winched the sailors out of the life raft.
B ______ The sailors inflated the life raft and jumped in.
C __1__ The boat struck an object in the sea.
D ______ The sailors fired two flares into the air.
E ______ The boat sank.
F ______ The beacon sent a signal to the satellite.
G ______ The beacon detached itself from the boat.
H ______ The rescue team saw the flares.

Reading 4 Read this news article and explain what the words below refer to.

SEVENTY or more kilometres from land, your boat strikes an unseen object and sinks quickly. You have no time to send a radio message. You jump into your life raft. You have flares in your life raft, but they are only visible from a distance of about 5 km. How do you send an emergency signal to the nearest rescue centre?

This happened to two sailors on 10 July this year. They were sailing in the Indian Ocean when their boat, the *Tiger*, struck a sharp object. The boat quickly sank 77 kilometres from the nearest land. They got into their life raft, but their radio was lost when the boat went down.

At 09.30 the coastguard received a signal from the boat's emergency beacon. The coastguard forwarded it to the rescue centre and by 11.00 (only 90 minutes later) the crew of the helicopter found the two sailors and winched them into the helicopter from the life raft. How was the emergency signal transmitted?

Fortunately, the *Tiger* was fitted with a 406 MHz free-floating beacon, which was linked to the Cospas-Sarsat satellite system. When the boat sank, the beacon automatically detached itself from the yacht and floated to the surface. There it switched on automatically and transmitted an emergency signal on the 406 MHz wavelength to the satellite. The satellite then forwarded the signal to the coastguard.

The free-floating beacon and the Cospas-Sarsat satellite system can increase the chances of saving lives in any air-sea rescue, in which the most important thing is to locate the survivors quickly.

1 They (line 14) *the two sailors*
2 it (line 25) ______
3 which (line 35) ______
4 itself (line 39) ______
5 There (line 41) ______
6 in which (line 52) ______

kilometres flares visible emergency signal coastguard beacon free-floating satellite automatically megahertz wavelength 03

5 Complete this incident report form.

INCIDENT REPORT FORM

Name of rescue helicopter pilot: *Ricardo Moussa* Date of rescue: ______

Name of boat: ______

Distance of boat from land: ______

Number of people rescued: ______

Time of first emergency signal: ______

Type of emergency beacon: ______

Time of rescue: ______

Method of rescue: ______

Speaking 6 Work in pairs. Take turns to be the rescue pilot and a safety officer. The safety officer interviews the pilot and asks questions based on the form.

Examples: *What's your name? When did the rescue take place?*

2 Transmission

Start here 1 Complete this description of how a satellite communication system works, using the correct form of the verbs in the box.

receive convert detach activate carry out transmit locate

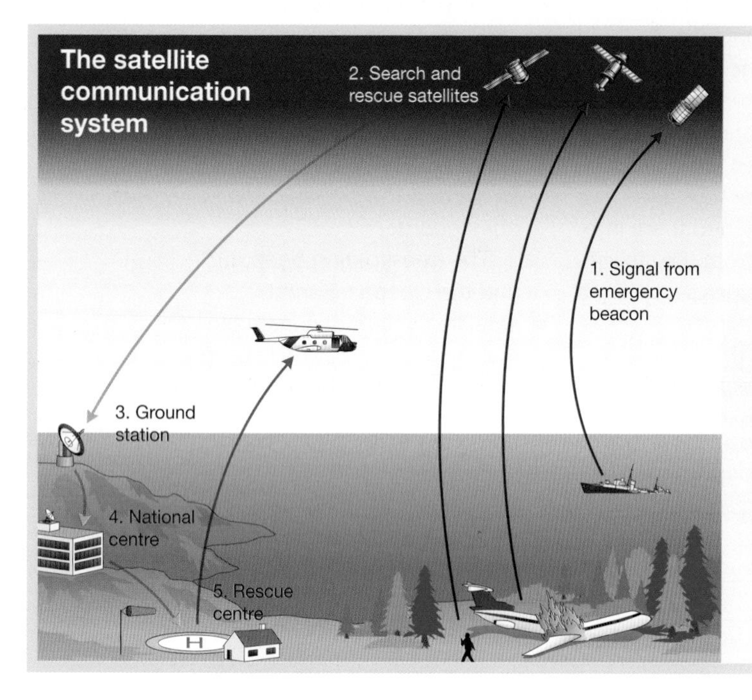

If a plane crashes, or a ship sinks, the survivors try to (1) ____________ their personal emergency beacons manually. In addition, an automatic beacon (2) ____________ itself from the plane or ship and switches on automatically. The beacon then (3) ____________ a signal to one or more satellites. The satellites (4) ____________ the beacon's transmission and then send the beacon's signal to their ground station. The ground station then processes the satellite signals (that is, it (5) ____________ the signals into useful data), and then passes on the data about the beacon to a national centre. The national centre forwards this data to the rescue centre nearest to the crashed plane or sinking ship. The rescue centre then (6) ____________ the beacon and sends out a rescue team, which then (7) ____________ the rescue.

Listening 2 04 Listen to this discussion and check your answers to 1.

Reading 3 Part of this text is missing. Write the letters of phrases A–G below in the correct spaces. Use the illustration in 1 to help you.

The Cospas-Sarsat system is an international search and rescue system which consists of a network of satellites in space, and control centres on Earth.

The components of the system are:

- radio beacons, which (1) ____________
- satellites, which (2) ____________
- ground stations, where (3) ____________
- national centres, from where (4) ____________
- rescue teams, who (5) ____________

The system uses two types of satellite:

- satellites in geostationary Earth orbit (GEO), which (6) ____________
- satellites in low-altitude Earth orbit (LEO), which (7) ____________

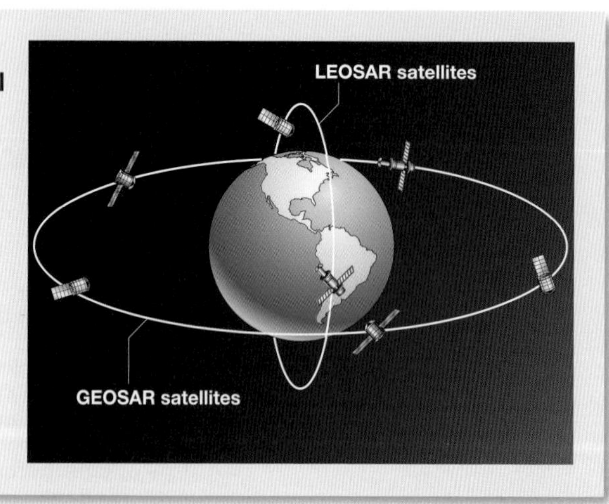

geo- = Earth
Geostationary satellites move at the same speed and in the same direction as the Earth. When we observe them, they seem to be *stationary* or not moving.

A are closer to the earth and cover polar regions.
B information about the emergency is sent to the rescue teams.
C are at a high altitude and cover a wide area.
D transmit 406 MHz signals in an emergency.
E signals from the satellites are processed.
F pick up the signals from the beacons.
G receive the information and carry out the search and rescue.

Language The relative pronoun (for example, *which*, *who*, *where*) is a useful way to join two sentences together.

Signals are transmitted to	the satellite. The satellite the satellite, which	then sends the signals to Earth.
The goods are taken to	the warehouse. Here the warehouse, where	they are stored safely.
This is	the city centre. From here the city centre, from where	roads lead in all directions.
Ricardo reports to	Waleed. Waleed Waleed, who	is the operations manager.

4 Join these pairs of sentences into single sentences. Use *which*, *where*, *from where* and *who* to replace the words in italics.

Example: *1 … to the satellite, from where …*

1. The beacon sends a signal to the satellite. *From here* the signal is transmitted to the ground station.
2. The rescue centre contacts the helicopter pilot. *He or she* then carries out the rescue.
3. The sailor activated his beacon. *This* sent a 406 MHz signal to the satellite.
4. The sailors were winched into the helicopter. *Here* they were given blankets and hot drinks.
5. The sailors were taken by helicopter to the rescue centre. *From here*, they were driven by ambulance to the nearest hospital.
6. Hundreds of survivors are saved every year by the Cospas-Sarsat system. *This* was first launched in 1982.

Speaking **5** Look at the table. Read out items a–h in full.

Example: *(a) (from) two to five kilograms*

Some facts and figures about the emergency beacon and the satellite system			
1	Radio frequency of beacon	a)	2–5 kg
2	Power (wattage) of beacon signal	b)	260 mm (h) x 102 mm (w) x 83 mm (d)
3	Length and frequency of beacon signal	c)	GME 203FF 18756
4	Dimensions	d)	35,000 km
5	Weight	e)	406 MHz
6	Operating range (temperature)	f)	-40°C–40°C
7	Model number	g)	5 W
8	Altitude of GEOSAR satellite	h)	0.5 sec every 50 sec

Task **6** Match items 1–8 with the correct items a–h in the table in 5.

Scanning **7** Practise your speed reading. Look for the information you need on the SPEED SEARCH pages (116–117). Try to be first to answer these questions.

1. When was the first Cospas-Sarsat satellite launched?
2. Which four countries started the Cospas-Sarsat system?
3. How many countries now operate the Cospas-Sarsat system?

3 Operation

Start here 1 Work in small groups. Study the diagram and discuss these questions.

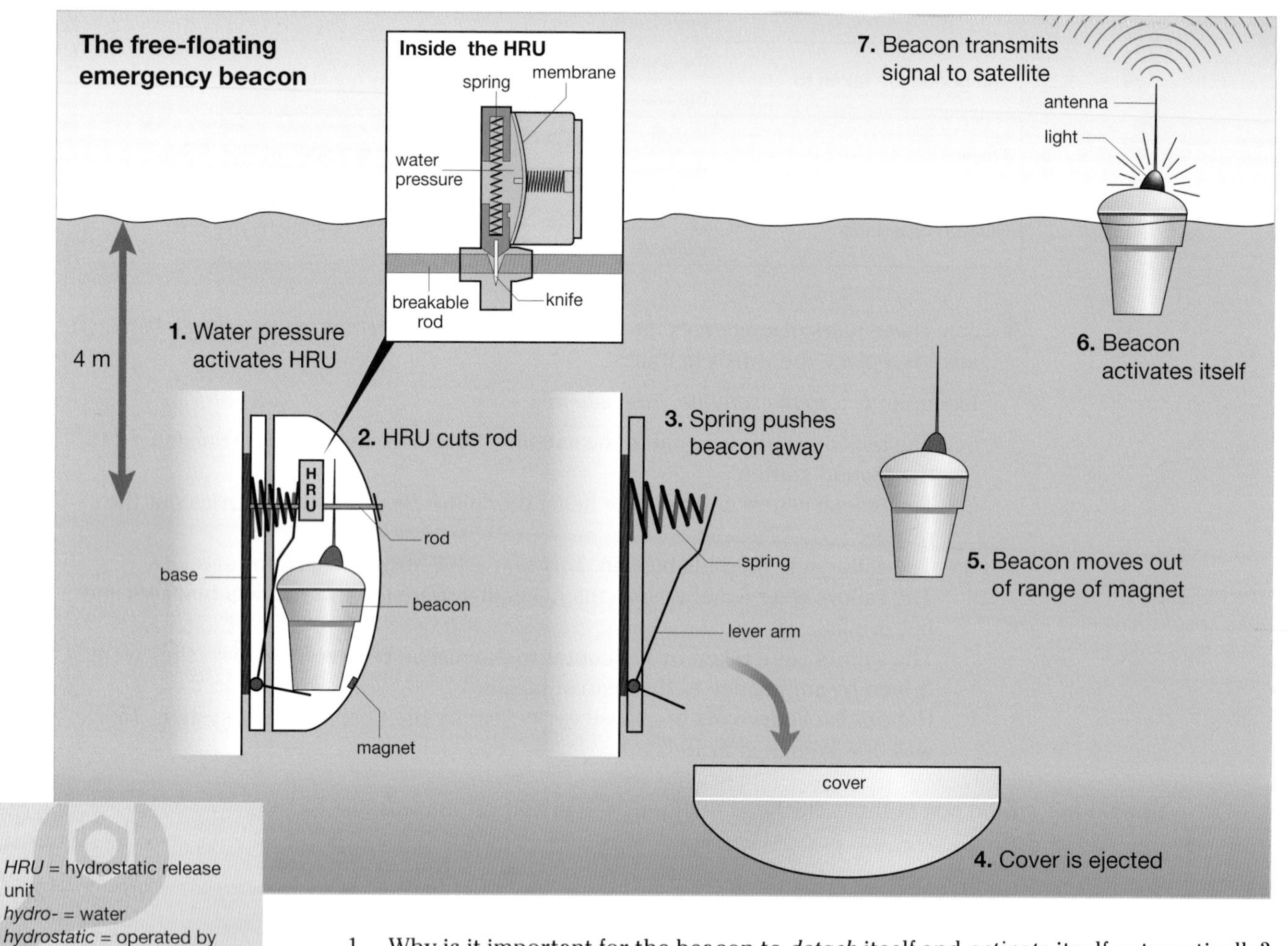

HRU = hydrostatic release unit
hydro- = water
hydrostatic = operated by water pressure

1 Why is it important for the beacon to *detach* itself and *activate* itself automatically?
2 How do you think it works?

Vocabulary 2 With your group, match synonyms a–e with the words in italics in 1–5.

1 the beacon *is submerged*
2 the rod breaks and this *releases* the cover
3 the cover *is ejected* from the base
4 the beacon moves *out of range of* the magnet
5 the beacon *activates* itself

a) *frees* it (*allows* it *to move away*)
b) *away from the force of*
c) *sinks under water*
d) *switches* itself *on*
e) *is pushed away*

Task 3 With your group, match questions 1–5 with answers a–e.

1 What does the rod do?
2 What makes the knife cut the rod?
3 After the knife has cut the rod, what pushes the cover away from the base?
4 What does the magnet do?
5 When the beacon floats away from the base, why does it switch on automatically?

a) Pressure from the spring and the lever arm.
b) Because it moves out of range of the magnet.
c) It fixes the cover to the base.
d) It prevents the beacon from switching on when it is inside the cover.
e) The pressure of the water and the force of the spring in the HRU.

Writing 4 With your group, write the *How It Works* section of an operating manual for the emergency beacon. Use all the information from the previous page. Complete the sentences to explain the seven stages in the diagram in 1.

Produce a single copy for your group. Each group member should work on different stages. Check and correct each other's work before you finalise the complete document.

Free-floating emergency beacon for Cospas-Sarsat rescue system

HOW IT WORKS

1 If the boat sinks, and the beacon is submerged below four metres of water, *the water pressure activates the HRU (hydrostatic release unit) automatically*.

2 The water pressure inside the HRU ...

3 This releases ...

4 The lever arm then ...

5 The beacon then ...

6 As a result, the beacon ...

7 When it reaches the surface ...

Vocabulary 5 Study the illustrations and supply the missing verbs in the instructions below.

ensure tear off touch remove slide place pull push

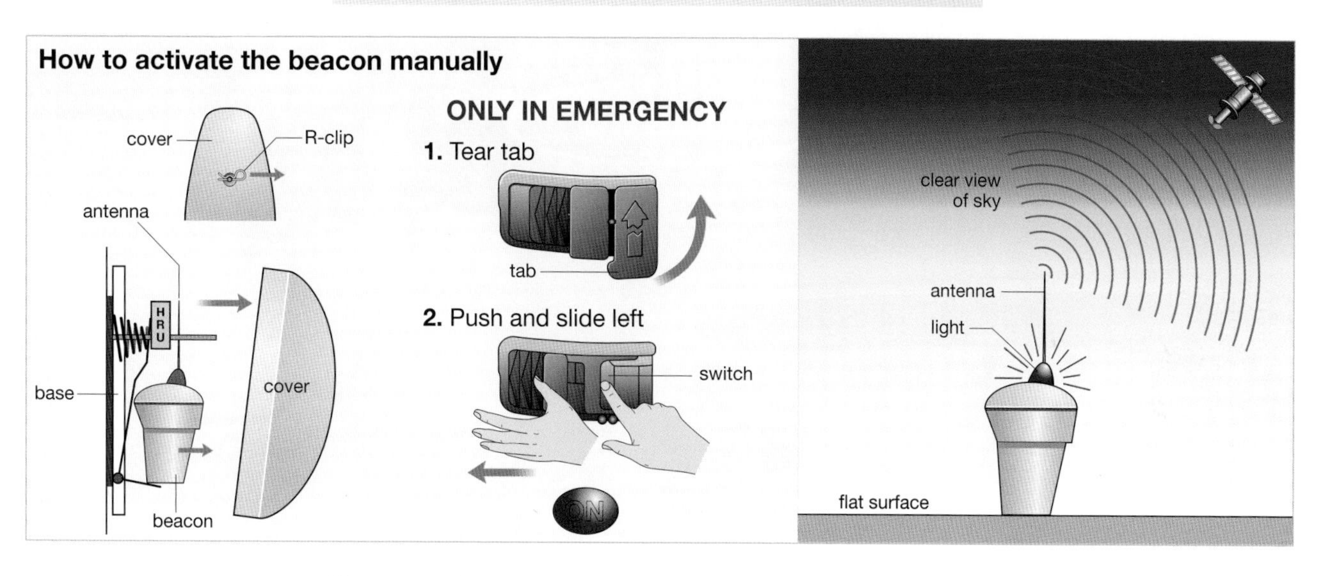

How to activate the emergency beacon manually

If the vessel is not sinking, but there is some immediate danger, you can activate the beacon manually. Follow these instructions:

1 ___Pull___ the R-clip.
2 ________ the cover and detach the beacon from its base.
3 ________ the tab. Underneath the tab is the switch.
4 ________ the switch down and ________ it to the left.
5 ________ the beacon on a flat surface and ________ that the antenna is upright. Check that the antenna has a clear view of the sky.
6 Do not ________ the antenna while it is transmitting.

Writing 6 Produce an operating manual with your group for a device you know about.

1 Agree on the device you want to write about.
2 Divide up the work. Each group member produces a different section of the operating manual: (1) *how it works*, (2) *operating instructions*, and (3) *labelled diagrams*.
3 Check each other's work, and then produce a single manual from the group.

2 Processes

1 Future shapes

Start here 1 Work in pairs. Look at the photos and discuss these questions.

1 Do we make these items from plastic now?
2 Do you think that we will make them from plastic in the future?

Listening 2 05 Listen to these five news reports. Match four of them with the pictures in 1.

A: News report ______ C: News report ______
B: News report ______ D: News report ______

What is the other report about? ________________

3 Listen again and write the report number under the correct heading.

designed but not yet manufactured	already manufactured and in use now	planned or expected in the future

Scanning 4 Practise your speed reading. Look for the information you need on the SPEED SEARCH pages (116–117). Try to be first to answer these questions.

Which plastic is used for making:

1 protective goggles ____________ ?
2 oars used in rowing boats ____________ ?
3 volleyball nets ____________ ?

Reading

5 Read this article, and write the letters of the paragraphs A–F which deal with these time frames.

1 the future _A_ ___ ___
2 a specified time in the past ___
3 an unspecified time in the past ___
4 the present ___

The future of plastics in aerospace engineering

A The world will be a very different place in the year 2035, and I believe that plastics will play an important role in that new world.

B In aerospace engineering, for example, it is probable that before 2035 they will make the fuselage (body) and wings of an aircraft entirely from plastics or plastic composites.

C However, it is unlikely that they will make the actual engine from plastics at any time in the future. And they certainly won't make one before 2035.

D Some manufacturers are trying right now to build aircraft fuselages from plastic composites.

E For example, one aircraft manufacturer has already designed a fuselage containing more than 50% carbon fibre (a plastic composite).

F They began the project three years ago, and they produced the drawings at the end of last year.

6 Tick the predictions below which are the same as the ones in the article in 5. Write the letter of the paragraph which includes the prediction.

1 It's possible that they'll build a plastic engine in the future. ☐ ___
2 It's likely that they'll construct a plastic wing before 2035. ☐ ___
3 They probably won't make a plastic engine before 2035. ☐ ___
4 They definitely won't manufacture a plastic engine before 2035. ☐ ___
5 They'll possibly make a plastic fuselage before 2035. ☐ ___
6 They probably won't build a plastic engine in the future. ☐ ___

Language

It's **certain** that they will They will **certainly** / definitely	make a plastic fuselage one day. not make a plastic engine before 2035.

It's **probable** / likely that they will They will **probably**	make a plastic fuselage before 2035. not make a plastic fuselage next year.

It's **possible** that they will They will **possibly**	make a plastic wing one day. not make a plastic wing before 2035.

7 Say each of the predictions in 6 in a different way with the same meaning.

Example: *1 They'll possibly build a plastic engine in the future.*

Speaking

8 Go round the class making predictions. Use dates and express either *certainty*, *probability* or *possibility* about each one.

Example: *I think that humans will probably reach Mars before 2040, but it's unlikely that they'll get to Saturn or Jupiter before then.*

2 Solid shapes

Start here 1 In pairs discuss the process illustrated in the diagrams, and answer the questions below.

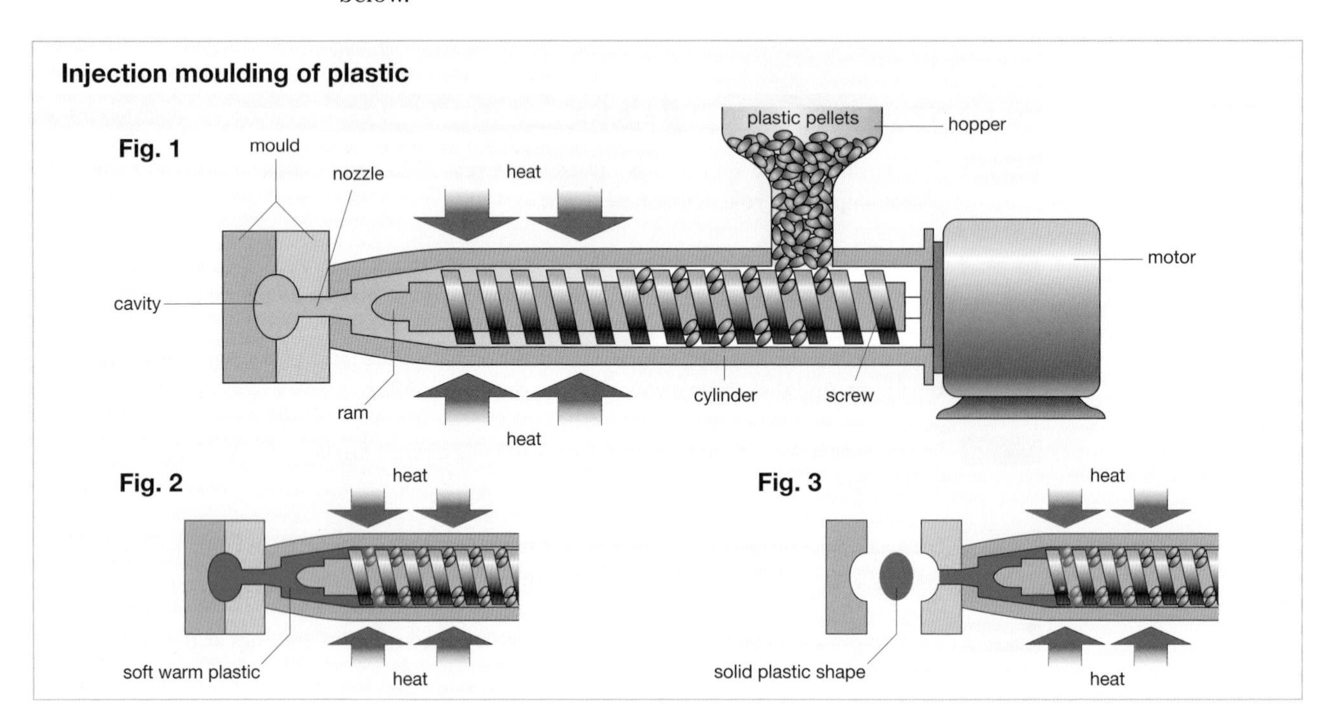

1 Why is heat used?
2 What is the function of the screw?
3 Which of the following items do you think are shaped using this process?

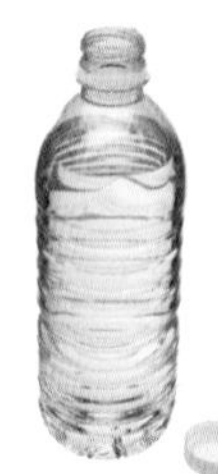

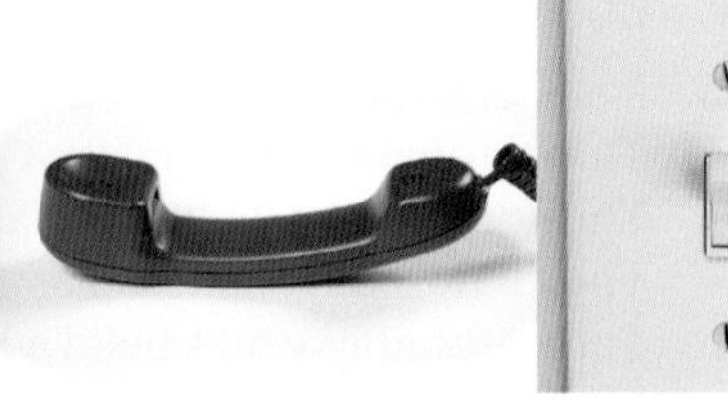
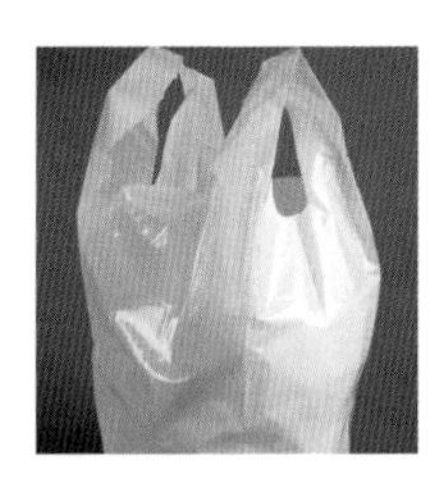

Reading 2 Rearrange these sentences into the correct order according to the diagrams in 1.

Injection moulding of plastic

A The mould opens and the cold, hard, solid plastic shape is ejected. ☐
B The screw stops rotating and then a ram in front of the screw moves straight forward. ☐
C Small pieces (or pellets) of plastic are fed from the hopper into a cylinder. 1
D The soft, warm plastic is pushed towards a nozzle by the ram. ☐
E The pellets are pushed along the cylinder by a rotating screw, and heated. ☐
F Inside the cavity, the plastic is cooled by the mould, and becomes hard. ☐
G The soft plastic is injected through the nozzle into a shaped cavity between the two halves of a mould. ☐

3 Check your answers to the three questions in 1.

Vocabulary 4 Find two words in 2 which contain the letters 'ject'. Which one means *thrown out* and which one means *pushed in*?

Language With an *active* verb, the *subject* = the *agent*. The subject carries out the action.

subject = agent	active verb	object
A rotating screw	pushes	the plastic pellets.

With a *passive* verb, the *subject* ≠ the *agent*. The subject does not carry out the action. The agent does the action to the subject.

subject	passive verb		agent
	be	past participle	
The plastic pellets	are	pushed	by a rotating screw.

5 Look at the diagram in 1 again and complete these sentences, using the active or passive forms of the verbs in brackets, as appropriate.

1. Plastic pellets ___________ (store) in a hopper at the top of the machine.
2. The pieces of plastic ___________ (transfer) from the hopper into a cylinder.
3. The plastic ___________ (propel) along the cylinder by a rotating screw.
4. The heaters around the cylinder ___________ (raise) the temperature of the plastic.
5. As a result, the soft, warm plastic softens and ___________ (flow) more easily.
6. The plastic ___________ (force) under pressure through a small nozzle.

Writing **6** Rewrite the paragraph below. Improve it by changing some (but not all) of the verbs to the passive form. Where appropriate, delete the agent. Make any other necessary changes. Begin some sentences with *First*, *Next*, *Now* and *Finally* as appropriate.

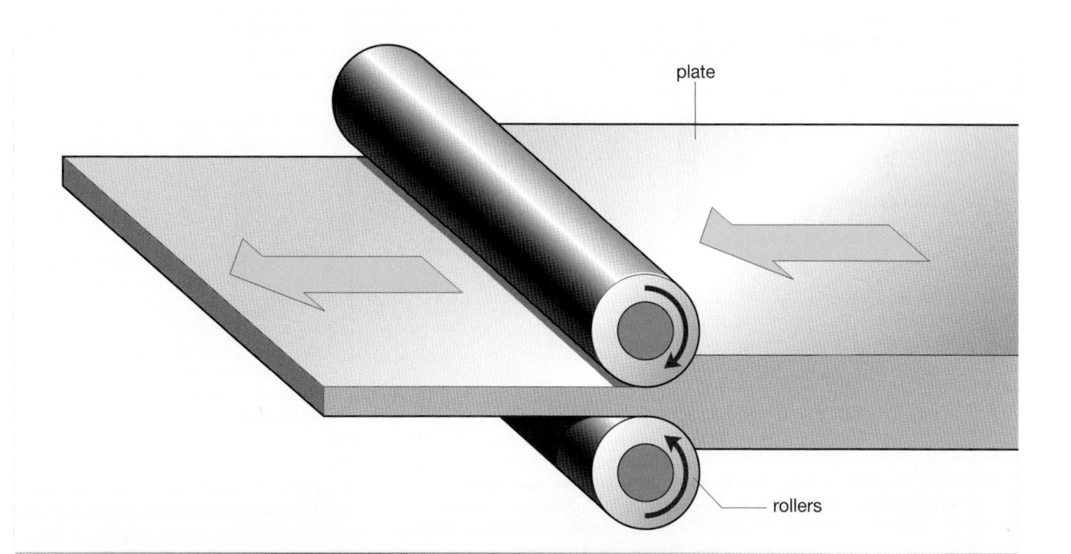

THE METAL-ROLLING PROCESS

Someone adjusts the gap between the rollers to the correct width. Someone switches on the motor, and the heavy rollers begin to rotate in opposite directions. A worker heats the metal plate. Then something pushes the hot metal plate through the gap between the rollers. As the hot metal moves between the rollers, the rollers compress it to a thinner shape. The metal comes out from the rollers in the form of a metal sheet. Someone then cools the metal sheet.

3 Hollow shapes

Start here

1 Work in pairs. How do you think the plastic items in the main picture were shaped? There's a clue in the two smaller photos.

Listening

2 With your partner, study the illustrations, and then rearrange the notes below into the best order for a talk on extrusion blow moulding.

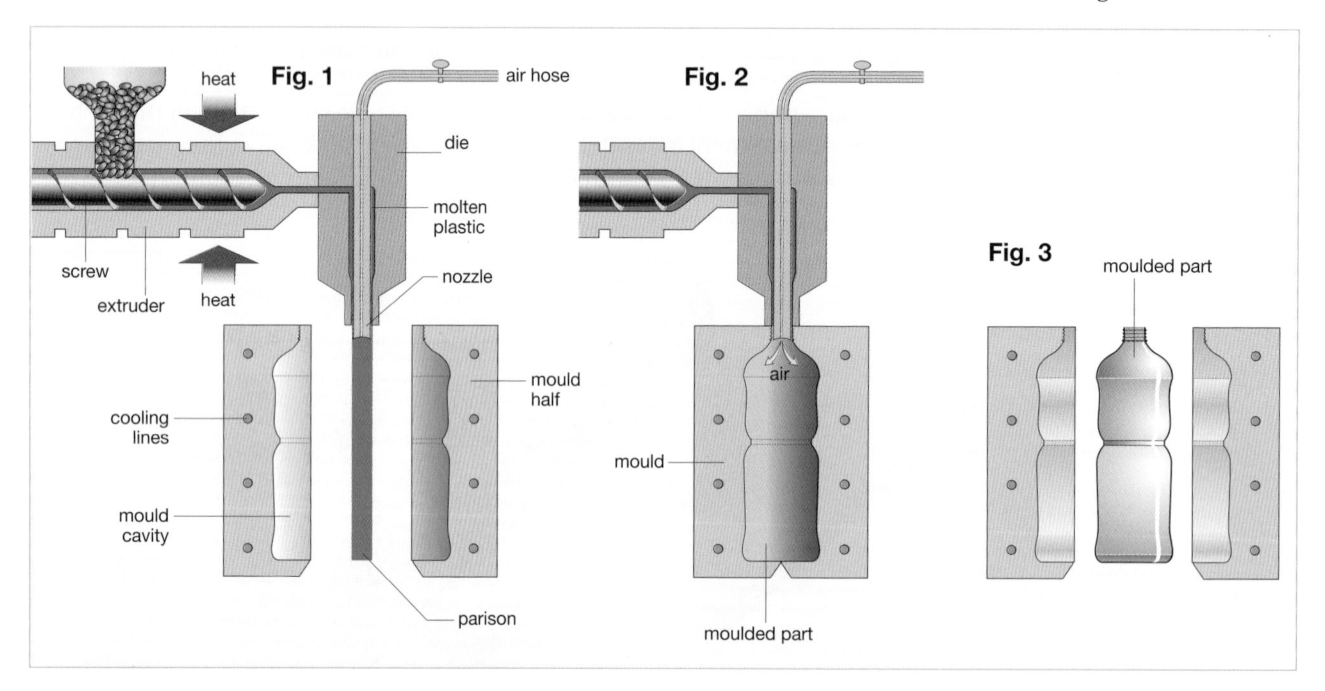

The extrusion process (SEE FIG. 1)

- movement of warm, soft molten polymer along cylinder ☐
- extrusion of molten polymer into mould ☐
- heating and melting of polymer pellets ☐
- transfer of polymer pellets from hopper to cylinder of extruder 1
- movement of cold polymer pellets along cylinder ☐
- rotation of screw ☐

The blow moulding process (SEE FIGS. 2 AND 3)

- cooling of plastic bottle shape ☐
- expansion of polymer to fit shape of mould ☐
- blowing of compressed air into molten polymer ☐
- ejection of plastic bottle from open mould ☐
- inflation of molten polymer by compressed air ☐
- closure of two halves of mould with molten polymer inside 7

polymer = plastic
molten = melted

3 ▶ 06 Listen to this talk and check the order of your notes.

4 Listen again and fill in the gaps with these phrases.

illustrates   shows   can be seen   is shown   is illustrated   can see

1 As you ____________ in Figure 1, there is an extruder at the top left …
2 As ____________ in Figure 1, there is a ninety-degree angle …
3 As Figure 1 ____________, the hot, soft plastic is extruded down …
4 Then, as Figure 2 ____________, the two halves of the mould close.
5 The second stage ____________ in Figure 2.
6 The third and final stage ____________ in Figure 3.

Vocabulary

5 Make a list like the one below. Write the first word from each note in 2 above under the noun column and write the related verb under the verb column.

noun	verb
movement	move

Language

6 Change the notes in 2 into full sentences, using the verbs from 5. Write them in the correct order. Use either active or passive verbs, as needed.

Examples:
1 The polymer pellets are transferred from the hopper to the cylinder.
2 The screw rotates.

Writing

7 Study the diagram and the notes below, and write an explanation of the process of pressure-die casting. Use *First*, *Then*, *Next* and *Finally*, and the passive where appropriate.

Begin: *First, some metal is heated until it melts. Next the molten metal …*

Pressure-die casting
- heat metal until it melts
- pour molten metal into chamber
- piston moves along chamber
- piston pushes molten metal under pressure into cavity
- cavity is between two halves of mould
- molten metal fills cavity
- metal cools and becomes solid
- open mould
- eject solid metal component from mould

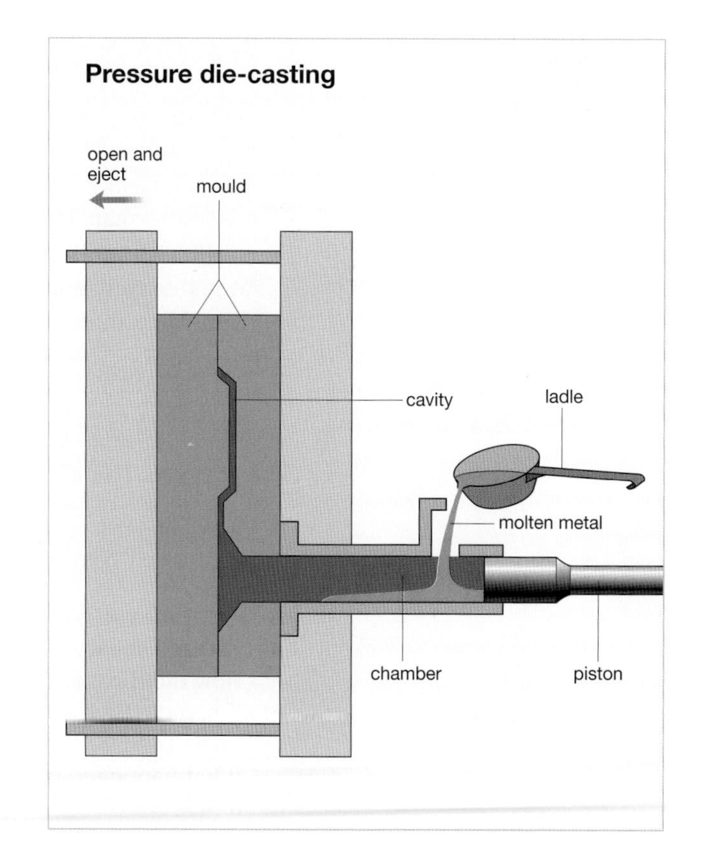

Speaking

8 Explain the process in 7 to the class, or to a partner. Do not look at your writing. Refer to the diagram where appropriate, using phrases from 4.

Example: *Next, as you can see in the diagram, the molten metal …*

Review Unit A

1 Write questions using the words in brackets to get these answers.

Example: *1 When did you transmit the message?*

1. We transmitted it yesterday evening. (When / the message)
2. It took place five miles along the North-South railway line. (Where / the accident)
3. I activated it by detaching it from its base. (How / beacon)
4. It struck the rock at 05.20 this morning. (When / the boat)
5. I inflated it because the crew were in immediate danger. (Why / life raft)
6. It took off at 14.38 this afternoon. (When / the aircraft)

2 What questions did the reporter ask to get the information for this news report? Use all these question words.

Where? When? How? Why? How long? How many?

At 14.20 yesterday afternoon, Air Traffic Control (ATC) received an emergency signal from the pilot of the plane. At the time of the signal, the plane was over the Atlantic, just off the coast of New Jersey. Three minutes later, ATC lost radio contact with the plane. The radio silence lasted for over ninety minutes. ATC contacted the coastguard to let them know the situation. Then a distress signal came from the plane's automatic emergency beacon. This signal provided data to the rescue team about the plane's location. The rescue team caught sight of the wrecked plane at 16.23. Then five minutes later, the team found the life rafts of the survivors and rescued them. More than two hundred passengers survived.

3 Complete the dialogue with the verbs in the box. Use all the verbs once each, in their correct form. There may be more than one possible answer for some gaps.

receive transmit detach pick up find out
locate send convert float activate

A: Can I ask you a few questions about the rescue of the sailors from the *Tiger*?
B: Sure, go ahead.
A: How did the automatic beacon work?
B: When the boat sank to four metres, the beacon (1) __________ itself from the boat. Then it (2) __________ up to the surface and (3) __________ itself.
A: What happened then?
B: It (4) __________ a signal to the satellite system.
A: So what time did the satellite (5) __________ the signal from the beacon?
B: I think it (6) __________ the signal at around 9.30.
A: Did the satellite (7) __________ the signal directly to the rescue team?
B: No, the signal passed through two control centres, which (8) __________ it into data. Then they sent the data to the rescue centre.
A: So when did the rescue team (9) __________ the sailors?
B: They (10) __________ where they were at about 11.00.

4 Complete these sentences, using *which, where* or *who*.

Someone is showing a visitor around the site of a telecoms project.

1 This is the office of the chief engineer, __________ runs the whole project.
2 The building over there is the warehouse, __________ all the parts are stored.
3 This structure here is the generator, __________ supplies electricity for the project.
4 And this is the central computer plant, __________ all the servers are located.
5 That tower is the main antenna, __________ transmits data to the satellite.
6 Those workshops are for the technicians, __________ do all the maintenance work.

5 Change this process description into a numbered set of instructions. Change every passive verb (in italics) into the imperative form.

Begin: *1 Turn the bike over, and place it upside down on the ground.*

How to remove a punctured bicycle tyre

First, the bike *is turned* over and *placed* upside down on the ground. Next, the nuts that hold the wheel to the frame *are loosened*. Next the wheel *is taken* out of the frame. The inner tube *is* then *deflated* completely by pressing down on the inner part of the valve. Next, two tyre levers *are used*. One lever *is pushed in* gently between the edge of the tyre and the rim of the wheel. The edge of the tyre *is* then *pulled out* over the rim. Now this lever *is left* in position, and the second lever *is pushed* between the tyre and the wheel rim. This second lever *is* then *slid* around the wheel rim under the tyre edge. Finally, the tyre and the inner tube *are detached* from the wheel rim.

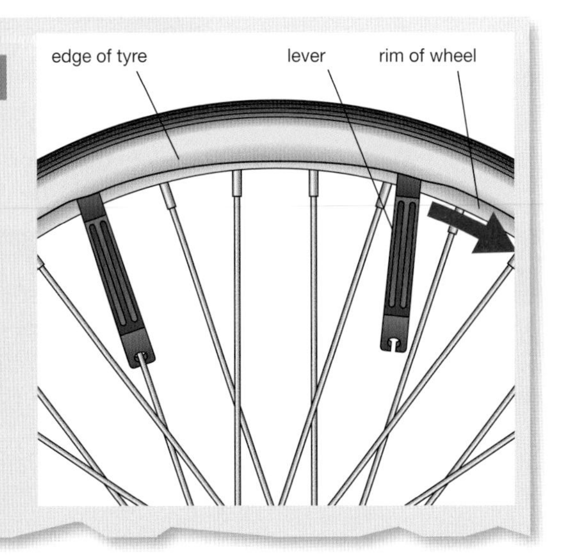

6 Change this set of instructions into a process description, using the passive. Change every imperative verb into the passive form. Use *First, Next, Now* and *Finally* as appropriate. Join two or more instructions together into the same sentence where appropriate.

Begin: *First, the cause of the puncture is located. The inner tube is submerged in water to locate the hole.*

How to repair a punctured inner tube

1 Locate the cause of the puncture.
2 Submerge the inner tube in water to locate the hole.
3 Pump a small amount of air into the tube to look for bubbles from the hole.
4 Mark the location of the hole.
5 Deflate the tube completely.
6 Rub the area around the hole with some rough material.
7 Spread a thin layer of glue around the hole.
8 Tear off the plastic backing from the patch.
9 Place the sticky side of the patch firmly on the tube.
10 Slide the tube back into the tyre.
11 Push the tyre and tube into the wheel rim.
12 Re-inflate the tyre.

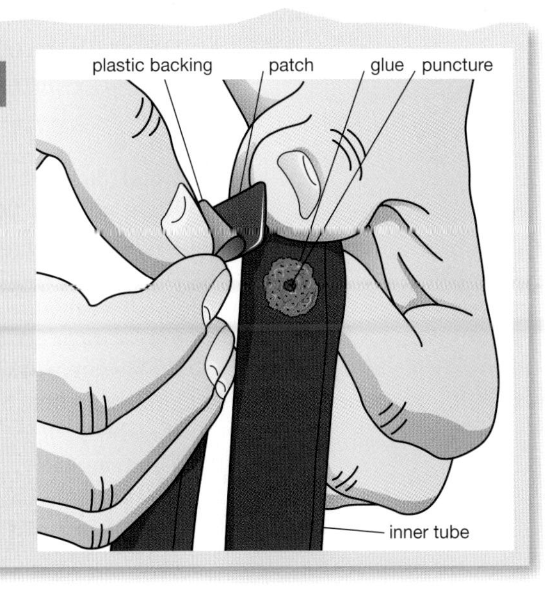

7 Work in pairs. Talk about what has just happened, what is happening now, and what you think will happen soon in these four situations.

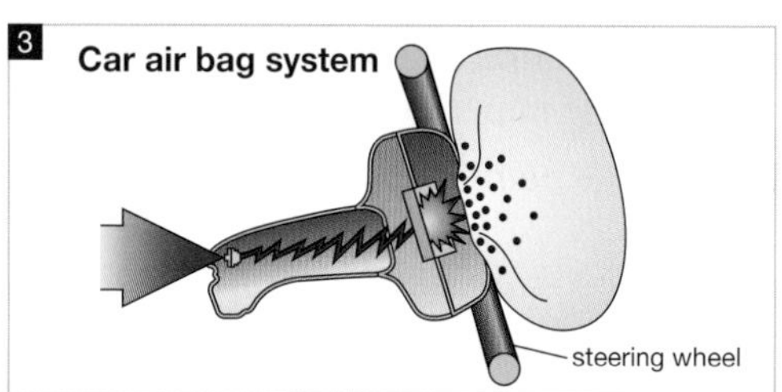

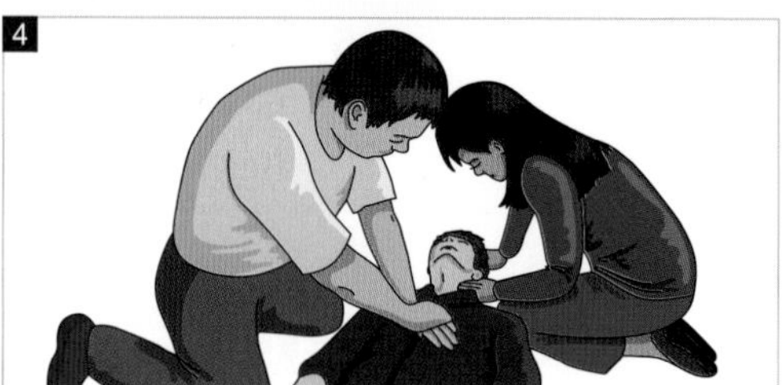

8 Complete the report with the correct form of the verb in brackets. Use *will*, the present simple, the present perfect and the present continuous.

Plastics in civil engineering

In the future, civil engineering projects such as roads and bridges (1) __________ (use) more and more plastics and composites.

Some of these developments (2) __________ (happen) right now. Many plastics manufacturers (3) __________ (design) plastic bridges at this very moment.

In fact, a plastic bridge already exists. A construction company (4) __________ (already / construct) the first European plastic road bridge in Hesse, Germany.

The engineers (5) __________ (install) the bridge on a single day early this year. This is how they did it. First, they (6) __________ (pre-fabricate) the bridge deck from a single piece of plastic five weeks before the installation date. Then they (7) __________ (transport) the deck as a single unit to the construction site. There they (8) __________ (fasten) it onto two steel supports. The total installation (9) __________ (take) less than a single day.

It is probable that the bridge (10) __________ (last) for up to 50 years without repairs, since the composite material (11) __________ (not / corrode).

9 Complete the dialogue below using the correct form of the verb in brackets.

A: OK, as it's the middle of March, could you give me a quick progress report on the project? What (1) __________ (you / do) up until now?

B: Well, we (2) __________ (design) the complete bridge at the end of January. And we (3) __________ (already / construct) the steel supports for the plastic deck.

A: Good. So what (4) __________ (you / do) at this moment?

B: Right now the polymer department (5) __________ (pre-fabricate) the plastic deck. That (6) __________ (be) ready before the end of March.

A: Right. So (7) __________ (you / manufacture) the other components yet?

B: No, we (8) __________ (not / make) them all yet. But the tool shop (9) __________ (make) them right now, and they (10) __________ (probably / finish) the work soon after the deck is ready.

A: OK, that's fine. So when (11) __________ (you / transport) the deck section to the site?

B: The team (12) __________ (take) it over there on installation day.

A: On the same day as the installation?

B: Yes, the team (13) __________ (carry) it there first thing, and they (14) __________ (attach) it to the supports before the end of the day.

10 Rewrite these sentences to give the same meaning using the words in brackets.

1 It's unlikely that engineers will invent a time machine in the future. (probably)
2 It's certain that they will manufacture an all-plastic car body some day. (definitely)
3 Astronauts probably won't be able to reach Mars in the next 20 years. (unlikely)
4 It's likely that sea levels will rise at least 0.2 metres in the next 50 years. (probably)
5 Beijing definitely won't make another Olympic bid in the next 20 years. (certain)
6 It's unlikely that an asteroid will strike the Earth in the next 100 years. (probable)

11 Rewrite these headings or captions changing the nouns in italics into verbs. Use the active or passive as appropriate

Example: *1 How aluminium sheet is extruded.*

1 The *extrusion* of aluminium sheet
2 The *injection* of diesel oil into the engine cylinder
3 The *propulsion* of land vehicles by the wind
4 The *rotation* of boat propellers in turbulent seas
5 The *ejection* of pilots from damaged planes
6 The *transfer* of survivors from the sea to the hospital
7 The *expansion* of steel beams at high temperatures
8 The *insertion* of agricultural sensors into the ground

12 Work in pairs. Discuss the four stages of this process. Make notes.

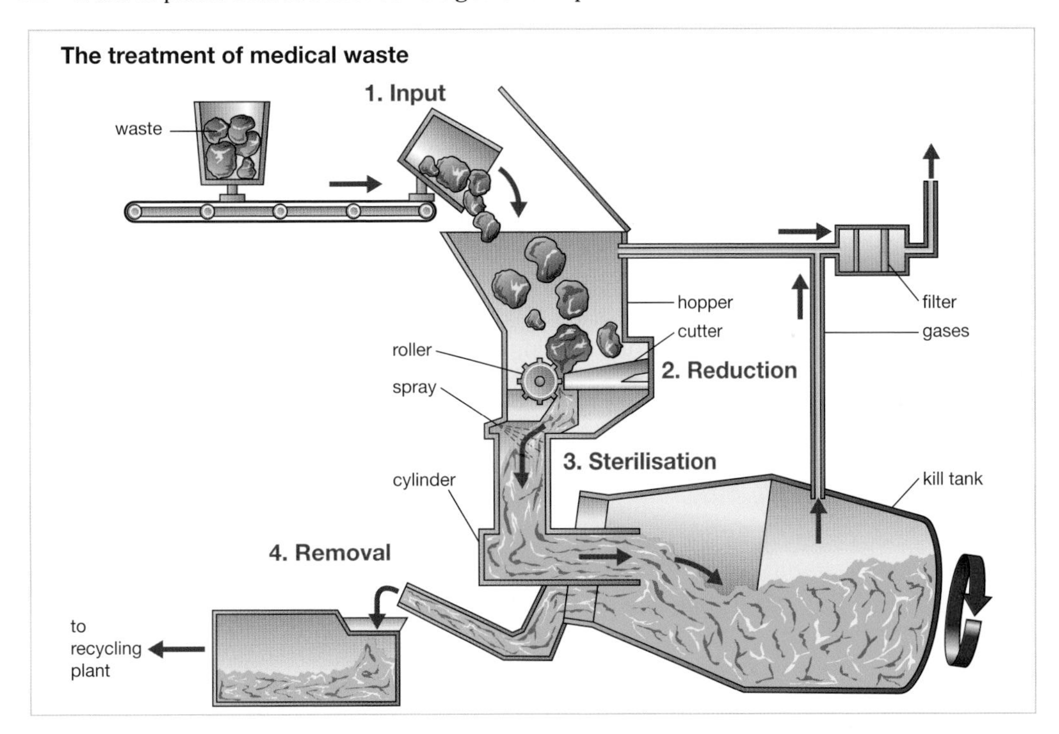

13 Write a short explanation of the process based on your notes.

14 Prepare and give a short talk on the process. As you move from one stage to another, make a reference to the diagram, using a different phrase each time.

Begin: *The diagram shows the four stages in the treatment of medical waste. As you can see in the diagram, in the first stage the waste material is carried to …*

Project **15** Research an important process in your own technical field and produce:

- a flow chart of the process
- a labelled diagram of the main equipment and its controls
- a description of how the process works
- a user's guide of how to operate the equipment / controls

3 Events

1 Conditions

Start here

1 How do the crew return to Earth? Discuss with a partner then briefly explain to the class.

Listening

2 07 Listen to this radio news report from the year 2020 and answer the questions. Use brief notes.

1 What is the purpose of the LAS?
2 What happens to the crew capsule after the LAS is activated?

3 Complete the first part of the news report using the correct form of the verbs in brackets. Check your answers in the audio script on page 119.

First, the news in brief. The new Ares moon rocket (1) ___________ (fail) to launch. The rocket (2) ___________ (crash) into the Indian Ocean. The crew capsule (3) ___________ (land) safely in the ocean.
Now, the news in detail. Six astronauts (4) ___________ (escape) death early this morning when their Orion crew capsule (5) ___________ (detach) itself safely from their Ares space rocket. The Ares rocket (6) ___________ (be / launched) at 5.05 this morning.

Language

Present perfect	Past simple
*The Ares rocket **has crashed**.*	*The Ares rocket **crashed** early this morning.*

4 Which verbs in 4 are (a) present perfect (b) past simple? Why?

5 Complete these radio news stories in the same way as in the news story in 3 (B = News in brief, D = News in detail).

1 B Four astronauts ___________ (land) on the Moon.
  D Their crew capsule ___________ (touch) down on the Moon at 03.44.
2 B Two military planes ___________ (collide) in mid air.
  D The two jet fighters ___________ (fly) into each other earlier today.
3 B A container ship ___________ (strike) an iceberg and ___________ (sink) in the Arctic Ocean.
  D The ship ___________ (run) into the iceberg early this morning, and finally ___________ (sink) at noon today.
4 B An oil rig in the Red Sea ___________ (burn down).
  D The Victory rig ___________ (catch) fire yesterday, and ___________ (collapse) early this morning.

Listening 6 08 In 2010, a scientist spoke about the invention of the LAS system. Listen to her explaining these slides and complete the sentences below.

(2010) WHAT WE HAVE NOW	(2020) WHAT WE DON'T HAVE (YET)
EJECTION SEATS FOR AIRCRAFT	**EJECTION CAPSULES FOR SPACECRAFT**
If an aircraft fails ... • pilot will activate ejection system • system will eject seat from plane	If a spacecraft failed ... • computer would activate LAS system • system would eject crew capsule

Slide 1: And if the pilot ____________ the ejection system, the system ____________ ____________ the seat, with the pilot, from the plane.
Slide 2: And if the computer ____________ the LAS system, the system ____________ ____________ the capsule, with the crew, from the spacecraft.

Language We use the **first conditional** when the situation is real or possible.

Condition		Result		
If an aircraft	fails,	the pilot	will activate	the ejection system.

We use the **second conditional** when the situation is imagined or impossible.

Condition		Result		
If a spacecraft	failed,	the computer	would activate	the LAS system.

7 Complete the inventor's statements about the LAS. Use the second conditional. Note that some gaps require two words.

1 If a spacecraft ____________ (have) an LAS system, the crew capsule *would have* (have) its own engines.
2 If the LAS ____________ (eject) the crew capsule, after some time the capsule's parachutes ____________ (open)
3 If the capsule's parachutes ____________ (open) correctly, the capsule ____________ (float) to Earth safely.
4 If the whole ejection system ____________ (work) correctly, the crew capsule ____________ (land) safely in the sea.

8 Complete this dialogue using the second conditional.

A: If you (1) *were* (be) head of space research at NASA, what (2) *would your research priorities be* (research priorities / be)?
B: Head of space research at NASA! That's impossible. I'm only a junior technician in a small aircraft company.
A: I know. So am I. But think about it: if you (3) ____________ (run) NASA's research programme, what (4) ____________ (you / do)?
B: Well ... I think I (5) ____________ (develop) the International Space Station into a world-class medical laboratory. How about you? What (6) ____________ (you / focus) on?
A: Well, if I (7) ____________ (be) the leader of NASA's research team, first I (8) ____________ (send) more spacecraft to Mars. And then if we (9) ____________ (find) water on Mars, I (10) ____________ (build) a city there.

Speaking 9 Work in pairs. Practise the dialogue in 8.

2 Sequence (1)

Start here 1 Work in small groups. Discuss how this system works and make notes. Use all the named parts in the diagram.

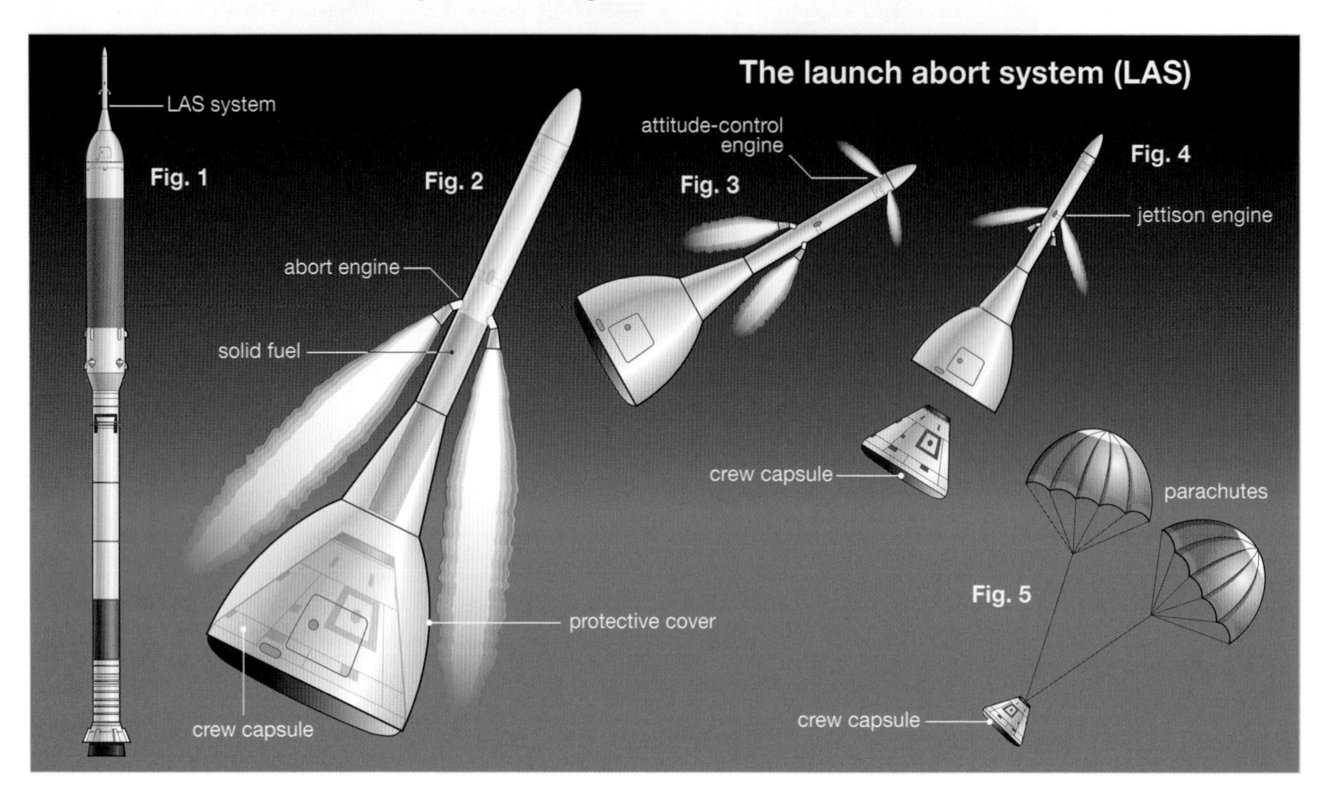

Listening 2 09 Listen to this presentation about the system and check your notes from 1.

3 Match the parts 1–4 with their functions a–d. Then listen again to check your answers.

Part	Function of part
1 jettison engine	a) stabilises the LAS (with the crew capsule) and makes it point in the right direction
2 protective cover	b) pushes the crew capsule away from the LAS
3 attitude-control engine	c) prevents hot exhaust from burning the crew
4 abort engine	d) pushes the LAS (with the crew capsule) away from the spacecraft

jettison engine protective cover attitude-control engine
abort engine launch abort system crew capsule stabilise
stabilisation orient orientation explosive bolts parachute 10

Speaking 4 Work in pairs. Ask and answer questions about the illustration in 1 and the table in 3. Use the second conditional.

Examples of questions: *What would the crew do if the jettison engine didn't work? If the crew capsule didn't have a protective cover, what would happen to the crew?*

Scanning 5 Practise your speed reading. Look for the information you need on the SPEED SEARCH pages (116–117). Try to be first to answer these questions.

1 How long is the LAS? ____________ m
2 What is the length of the crew capsule? ____________ m
3 How quickly can the LAS reach its maximum speed? ____________ seconds
4 What pressure can astronauts experience during ejection? ____________ Gs

1 G = the force of the Earth's gravity

Reading 6 Read this text and put the events below in the correct order.

How the launch abort system (LAS) of the Ares Moon Spacecraft works

If the rocket is launched and then ascends to more than 100,000 metres, the launch abort system (LAS) is automatically disabled. However, if there is a problem with the launch within 100,000 metres, the LAS is activated. This is how it works.

After detecting the fault, the rocket's computer system activates the abort engine. This thrusts the LAS (attached to the crew capsule) away from the rocket.

When the abort engine has burnt out, the LAS continues to move away from the falling rocket. Now it is guided by the attitude-control engine at the tip of the LAS. The engine can spin the LAS around in any direction to stabilise it and orient it.

As soon as the LAS has reached the correct orientation, the explosive bolts detonate. Immediately afterwards, the jettison engine fires and pushes the crew capsule away from the LAS. Once the capsule has reached a safe altitude, parachutes open up and allow the capsule to float down into the ocean.

A parachutes open ☐
B capsule separates from LAS ☐
C explosive bolts detonate ☐
D jettison engine fires ☐
E abort engine burns out ☐
F LAS is ejected from rocket ☐
G attitude-control engine operates ☐
H abort engine is activated [1]

Language

<table>
<tr><th colspan="4">First action</th><th>Second action</th></tr>
<tr><td colspan="4">The capsule reaches a safe altitude.</td><td>Then the parachutes open up.</td></tr>
<tr><td>After</td><td rowspan="4">the capsule</td><td rowspan="2">reaches</td><td rowspan="4">a safe altitude,</td><td rowspan="4">the parachutes open up.</td></tr>
<tr><td>Once</td></tr>
<tr><td>When</td><td rowspan="2">has reached</td></tr>
<tr><td>As soon as</td></tr>
</table>

If both clauses have the same subject, we can use this form.

<table>
<tr><th colspan="3">First action</th><th>Second action</th></tr>
<tr><td colspan="3">The computer detects a problem.</td><td>Then it (= the computer) activates the abort engine.</td></tr>
<tr><td>After</td><td>detecting</td><td>a problem,</td><td>the computer activates the abort engine.</td></tr>
</table>

7 Underline examples of the above forms in the text in 6.

8 Make full sentences from these notes. Use as many forms from the language box as you can.

Example: 1 After checking all systems, the computer begins the countdown.

1 computer checks all systems ➔ computer begins countdown
2 countdown ends ➔ rocket is launched
3 computer activates abort engine ➔ LAS is ejected
4 abort engine fires for three seconds ➔ abort engine burns out
5 crew capsule separates from LAS ➔ parachutes open
6 parachutes open ➔ parachutes lower capsule gently into sea

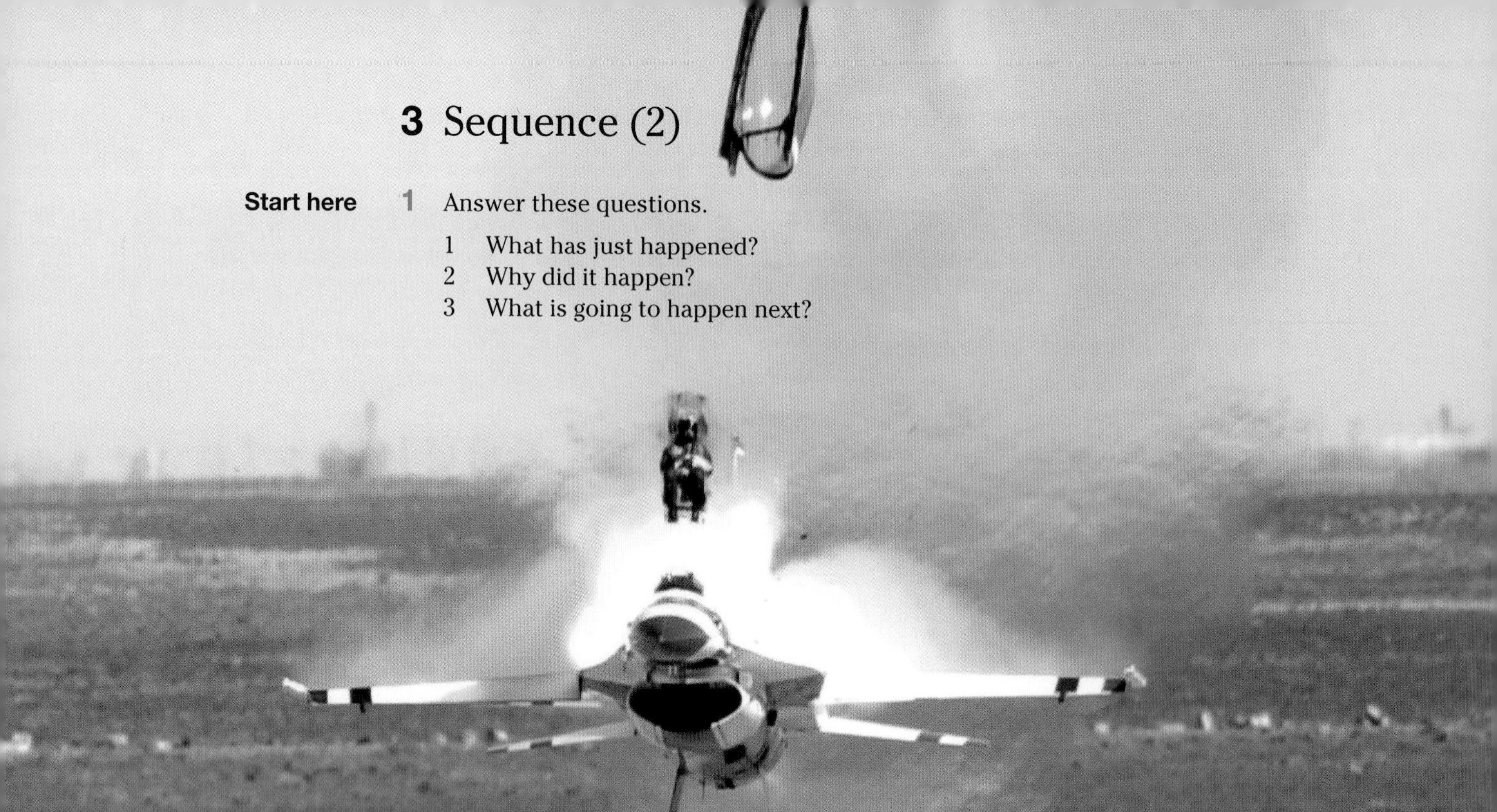

3 Sequence (2)

Start here **1** Answer these questions.

1 What has just happened?
2 Why did it happen?
3 What is going to happen next?

Task **2** Work in pairs. Put these illustrations in the best order to explain how the ejection seat system works.

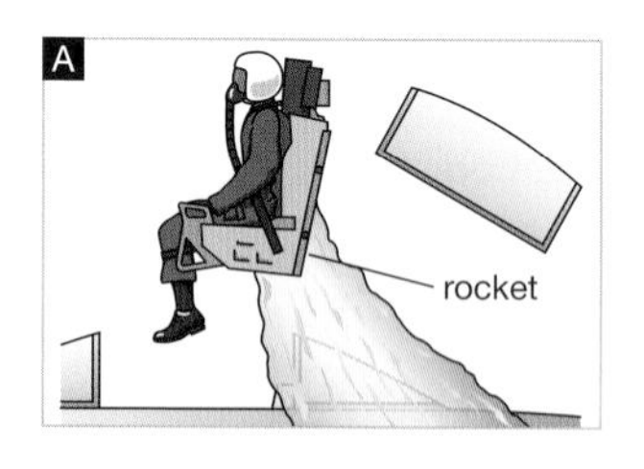

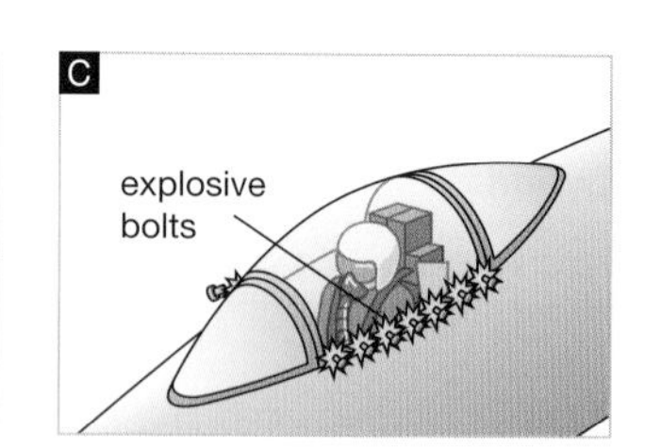

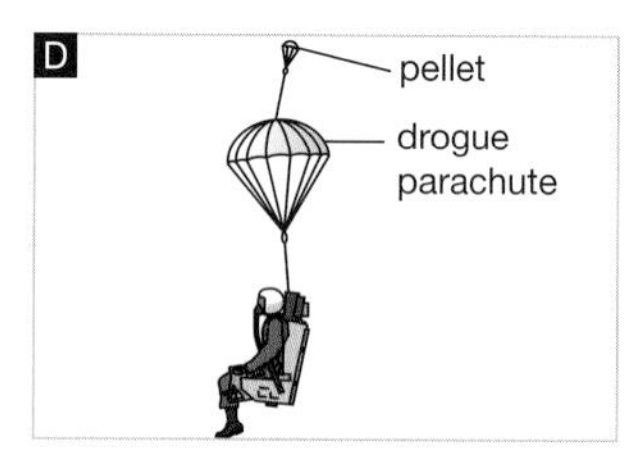

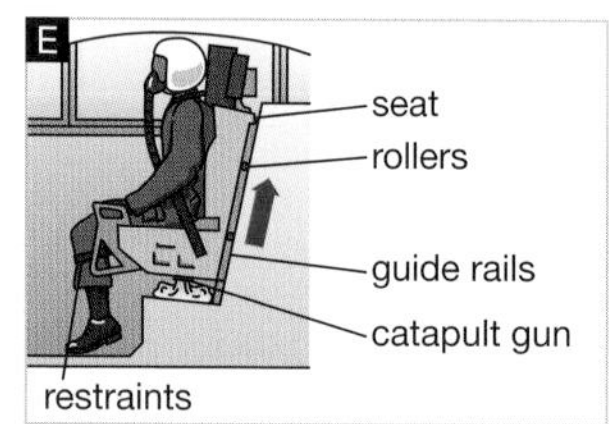

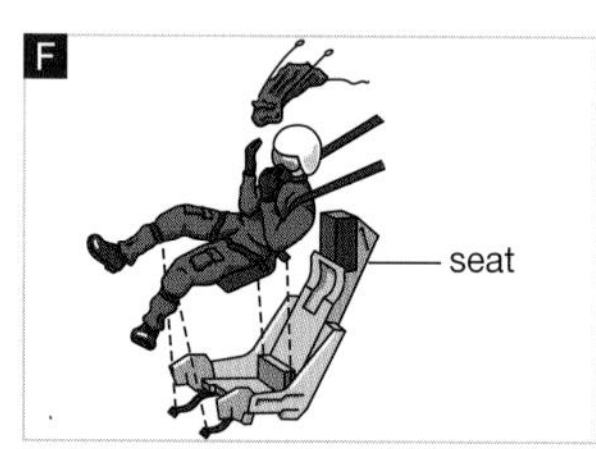

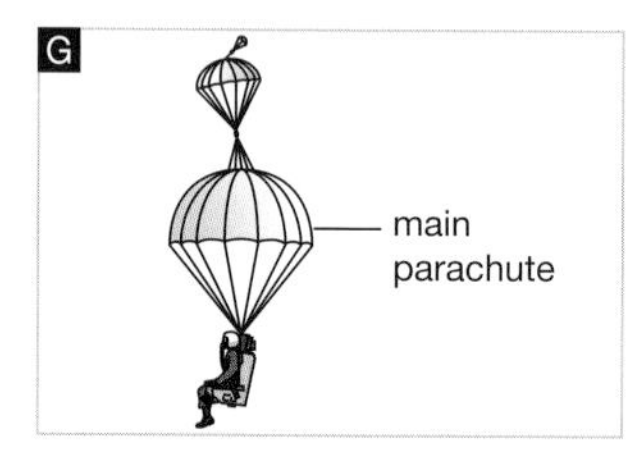

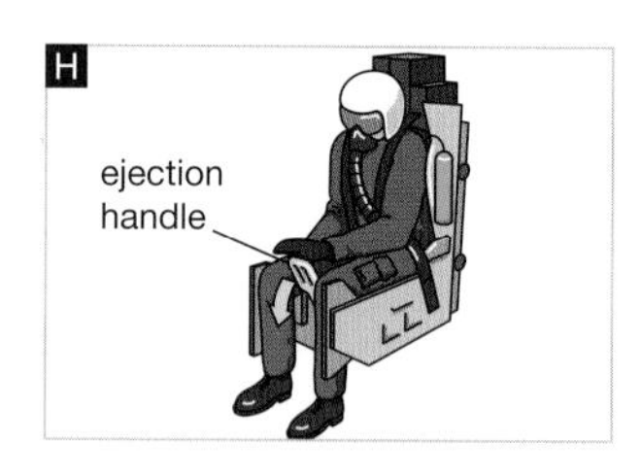

Vocabulary **3** Match the phrases 1–6 with the same or similar meanings a–f.

1 eject the seat	a) stop it from moving
2 stabilise the rocket	b) go down
3 deploy the parachute	c) make it steady
4 restrain the body	d) throw it out
5 descend	e) open it and use it
6 ascend	f) go up

4 Complete these nouns with the correct endings in the box.

-ment -ation -t -ion

1 deploy-	3 stabilis-	5 activ-
2 eject-	4 orient-	6 restrain-

Task **5** Work in pairs. Ask each other about the missing information for the ejection sequence and complete your chart.

Student A:

1 Ask Student B questions to help you complete your chart below.
2 Use the text on page 111 to help you answer Student B's questions.

The first part of the ejection sequence (0–0.15 seconds)

1 The ejection seat system is activated by pulling the ejection ________.	0 sec
2 Next, the bolts ________. This action releases the ________.	
3 Then, the ________ flies away from the ________.	
4 After this, the ________ is fired to make the ________ rise.	
5 The rollers of the seat move along the ________.	
6 The ________ prevent your legs from moving away from the seat.	
7 The seat ________ the cockpit before the rocket engine fires.	0.15 sec
8 As a result, the seat is propelled ________ from the plane.	

Student B:

1 Use the text on page 115 to help you answer Student A's questions.
2 Ask Student A questions to help you complete your chart, which is on page 109.

Writing **6** Read this article about racing cars from the year 2025. Work in a small group to carry out the task below.

Like any other sport in the year 2025, professional racing has changed. It's become bigger, faster, and more dangerous. Racing cars now often reach an average of over 600 kph. All racing cars are equipped with ejection seats, and use radar and sensors to prevent the drivers from hitting bridges or other obstructions

You are a design team working in the year 2025. Design an ejection seat system for racing cars so that the driver can escape safely just before a crash. Think about possible problems (such as bridges, or spectators) and how your design would solve them. Use the language in 5 to help you.

7 With your group, write a description of how your ejection system works.

Speaking **8** Explain your group's design to the rest of the class, or to another group.

4 Careers

1 Engineer

Start here

1 Discuss these questions in pairs.

1 Do you have your own blog page?
2 If you do, what information have you put on it?
3 If not, what information would you put on it if you had one?

2 Look at Hans's blog and complete his mini-profile below.

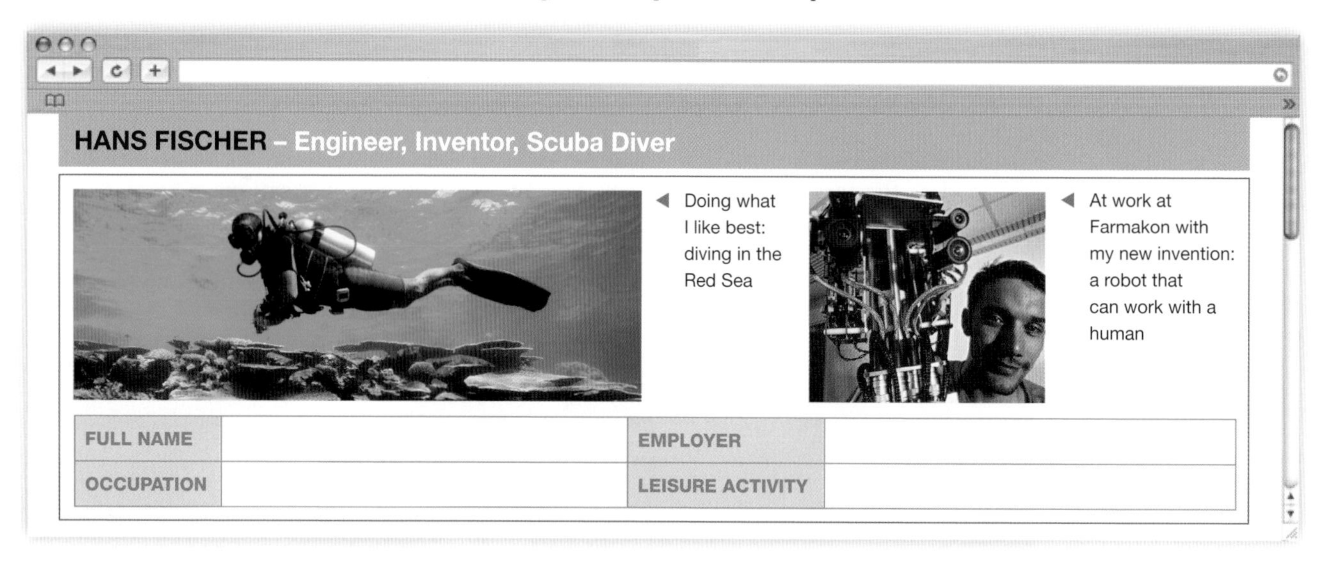

Listening

3 11 Listen to this interview with Hans and answer the questions.

1 Where is Hans from?
2 Where is he working at the moment?
3 For how many years has he worked there in total?
4 What does he do every day?

4 Listen again, and correct five mistakes in Hans's profile.

Profile of Hans Fischer

Company	Farmakon International GmbH
First job in company	Apprentice technician
Age joined company	18
Length of apprenticeship	Four years
Current job title	Mechatronics Engineer
Job routine	I design and build robots for the packing lines. Then I test them. I work with robots every day.
Years in current job	Eight years
Qualifications	Bachelor of Engineering degree in Robotics
Details of invention	I've designed a robot that can work safely with a human.
Current activity	At the moment I'm piloting the robot on a packing line.
Next step	Next month I'm going to build another robot for a different line.
Career plans	I'm doing a Masters degree next year.

Language

Present simple	routine activity	*I **work** with robots every day.*
Present continuous	current activity	*At the moment **I'm piloting** the robot.*
	planned activity	***I'm doing** a Masters degree next year.*
going to	intended activity	*Next month **I'm going to build** another robot.*

Speaking 5 Work in pairs. Ask and answer these questions.

1 Where are you working or studying at the moment?
2 What project or task are you working on at the moment?
3 What are you going to do after you complete your current project or task?
4 What is your job description? What do you do in your job (or studies)?
5 What do you normally do each day of the week? Select the most important.
6 What are you doing next in your career, after you finish studying / working here?

Reading 6 Complete Hans's CV. Check your answers in the audio script on page 120.

CURRICULUM VITAE: Herr Hans Fischer

PERSONAL INFORMATION

NAME: FISCHER Hans
DATE OF BIRTH: 18/04/1986
NATIONALITY: German

WORK EXPERIENCE

DATES: July 2012–present
POSITION: (1) ________
RESPONSIBILITIES: (2) ________
EMPLOYER: (3) ________
TYPE OF BUSINESS: (4) ________

EDUCATION

DATES: October 2008–July 2012
QUALIFICATION: (5) ________
MAIN SUBJECT: (6) ________
INSTITUTION: (7) ________

INDUSTRIAL TRAINING

DATES: July 2002–July 2005
TRAINING: (8) ________
INSTITUTION: Farmakon International GmbH, Munich

PERSONAL SKILLS AND COMPETENCES

SOCIAL AND ORGANISATIONAL: Captain of Farmakon Apprentices Football Club 2003–2004

apprentice technician robotics mechatronics
pharmaceutical curriculum vitae CV responsibilities
employer business qualification institution competences 12

Writing 7 You are Hans and you want to apply for the job in this advertisement in *Robotics* magazine. Write a covering letter to send with your CV.

2 Inventor

Start here

1 Work in pairs. Study the diagram and discuss these questions. Make notes of your answers.

1. What do you think these two devices are used for?
2. What appear to be the main differences between the two devices?
3. What do you think was wrong with the standard design?

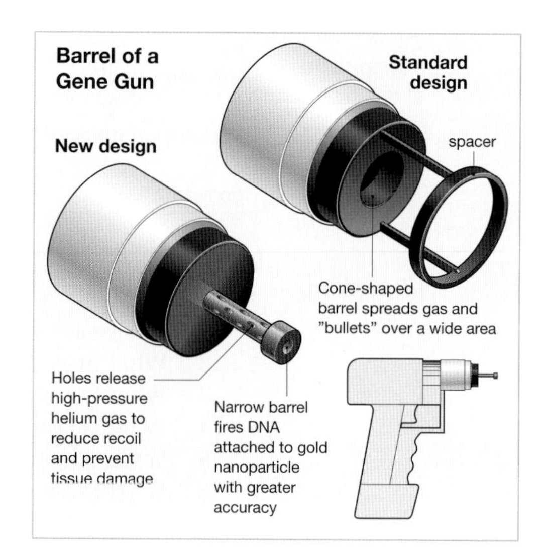

Reading

2 Read this article and check your answers to 1.

FROM technician TO gene gun expert

John O'Brien produced the prototype of his modified gene gun while he was a laboratory technician. Over the last few years, he has become an international expert in gene gun technology. He has given lectures and written research papers on different applications of the gene gun.

A gene gun is a tool which inserts genetic information such as DNA into a cell. It is hand held and **roughly** the size of a hairdryer. The standard gene gun was originally developed to **insert** material into plants. It is powered by a **pulse** of helium gas, which pushes its genetic bullet easily through the cell **membrane** and into the cell.

Scientists needed a gene gun for research into animal and human brain cells. But the standard gene gun used for plants had one major problem: it was too **inaccurate** for brain cells. It would hit the target, but it also hit everything around it at the same time.

In the standard gene gun, the barrel is wide, and conical in shape. This spreads the gas and 'bullets' over a wide area, which causes tissue damage around the target cell. The high-pressure gas causes a **recoil**, which creates more inaccuracy.

John worked to **reduce** the gas pressure to improve the gene gun's accuracy and prevent tissue damage. For some time, he could not find a solution to the problem.

Then John made his **breakthrough** when he met a former policeman who showed him how a machine gun works.

The barrel of a machine gun contains a number of holes which are set at an angle to the line of the barrel. This angle allows the air to escape from the holes more quickly.

John decided to put holes in the barrel of the gene gun. In the **modified** gene gun, the **external** barrel is much narrower than the standard gun (5 mm diameter instead of 40 mm) and is straight, not conical. This fires the DNA with greater accuracy.

The barrel is made of brass and stainless steel, and has about 20 holes, which allow the helium gas to escape. This reduces recoil and **prevents** tissue damage. John experimented until he found the **optimum** angle for the holes (about 30 degrees) for **minimum** tissue damage. A final **difference** is that, unlike the standard gene gun, the barrel of the modified model has no spacer, and therefore goes closer to the tissue.

John's modified gene gun is so accurate that biology labs around the world are currently using more than 150 of them.

3 Read the article again. Fill in the gaps and delete the false information in this chart.

Key differences between the standard and modified gene gun

	Standard	Modified
Diameter of barrel	______ mm	______ mm
Spacer?	yes / no	yes / no
Holes in barrel?	yes / no	yes / no
Shape of barrel	conical / straight	conical / straight
Recoil	high / reduced	high / reduced

Vocabulary

4 Find the synonyms in bold in the text in 2 of these words and phrases.

1 powerful backwards movement
2 changed and improved
3 short burst of power
4 sudden new idea
5 best
6 approximately
7 outside
8 thin layer of skin

5 Find the opposites in bold in the text in 2 of these words and phrases.

1 maximum
2 increase (vb)
3 exact
4 similarity
5 extract (vb)
6 allows

Language

comparative adjective + *than*	*The modified barrel is **narrower than** the standard one.*
***more / less* + adjective / noun**	*It's also **more accurate** and has **less recoil**.*
but / while / whereas	*It's straight, **but / while / whereas** the other one is conical.*

6 Make sentences comparing the two gene guns.

Examples: The modified design has a narrower barrel than the standard one. The standard design has a spacer, whereas the modified one doesn't.

Speaking

7 Ask questions to get these answers about the modified gene gun. Check the information in the text in 2 before you ask the questions.

1 It's for inserting genetic information into a cell.
2 It's about the size of a hairdryer.
3 By a pulse of helium gas.
4 Brass and stainless steel.
5 It's only 5 mm in diameter.
6 To allow the helium gas to escape and reduce recoil.
7 It's much narrower.
8 The modified barrel is straight, whereas the standard one is conical.

8 Work in pairs. Take turns to role-play the parts of a reporter and the inventor, John O'Brien. The reporter asks him about the modified gene gun and how it differs from the standard design.

Ask about:

(a) the modified gene gun itself, for example, its *shape*, *dimensions*, *main components*, *function or purpose*, *materials* and *method of operation*.
(b) ways in which the modified design is *different from*, or *compares with*, the standard design.

3 Interview

Start here **1** Put the advice about job interviews below under the correct headings. Write the numbers 1–10 in the table.

Before the interview DO ...	During the interview DON'T ...	During the interview DO ...

1 show your knowledge about the company
2 prepare a list of questions you would like to ask the interviewer
3 act as if you're not really interested in the job
4 answer only *Yes* or *No*
5 find out about the company and the job
6 talk negatively about your previous employer
7 be positive and honest about yourself
8 prepare a list of the questions you think the interviewer will ask you
9 ask questions about the job
10 check the job advert and think how your CV matches what they want

Listening **2** 13 Listen to part of an interview Reme Gomez has for a new job. Tick the advice in 1 which she follows.

3 Listen again and complete the interviewer's notes of the interview.

Name – Reme Gomez
Present job – ______________
Years in present job – ______________
Years in apprenticeship – ______________
Qualification – ______________
When gained qualification – ______________
Studying for diploma (part-time) now
Technical skills – (1) accurate ______________ *(2) CAD/CAM*
Personal skills – works hard, punctual and ______________
Interpersonal skills – willing to learn, and good ______________

Vocabulary **4** Match words or phrases 1–8 with the same or similar meanings a–h.

1	competence	a)	pay
2	qualification	b)	person who works, or is employed
3	experience	c)	previous work
4	salary	d)	for example, a company car or a pension
5	job title	e)	the person or company that employs you
6	benefit	f)	skill
7	employer	g)	for example, a degree or diploma
8	employee	h)	position (or post)

5 Complete the questions. Use words from 1–8 in 4.

1 What are your ______? I have a diploma and a degree in Engineering.
2 Who is ______? I work for VW.
3 What ______? It's 'Automotive Technician'.
4 What ______? I have good technical and communication skills.
5 What ______ in your job? I get a company car now that I'm a manager.

Language Reme's career at MultiPlastics

apprentice 2008 → 2011
junior technician 2011 → present day

Present perfect	*for*	*Reme's worked at MultiPlastics for seven years.*
	since	*She's had a job as a junior technician since 2011.*
	from ... until now	*She's been at MultiPlastics from 2008 until now.*

Past simple	*for*	*Reme was an apprentice at MultiPlastics for three years.*
	from ... to / until	*She worked as an apprentice from 2008 to / until 2011.*
	ago	*She joined MultiPlastics seven years ago.*

6 Supply the questions for Reme's answers.

1 How long ______? I've been a junior technician for the last four years.
2 When ______? I became an apprentice seven years ago.
3 How long ______? I was an apprentice for three years.
4 How many years ______? I've worked for MultiPlastics for about seven years.
5 How long ______? I studied for my certificate for two years part-time.
6 When ______? I received my certificate five years ago.
7 How long ______? I've been a part-time diploma student for two years now.
8 How long ______? I think MultiPlastics has been in business since 1995.

Speaking

7 Work in pairs, A and B. Take turns to interview each other. Look back at Hans's CV on page 27 and ask and answer questions about it.

Student A. You are Hans. Answer questions about your CV.
Student B. You are the interviewer. Ask Hans questions about his CV.

Scanning

8 Practise your speed reading. Look for the information you need on the SPEED SEARCH pages (116–117). Try to be first to answer these questions.

Search for an advertisement for the job of Electrical Design Manager.

1 What qualifications are required for the job?
2 What software must the manager be skilled in using?
3 Where will the manager need to travel in his/ her job?
4 What will the basic salary (minus bonuses) be?

Task

9 Imagine you are ten years older. Write a short version of the CV you would like to have then.

10 Note down some details of a job you would like to apply for in ten years' time.

11 Prepare for a job interview. Write notes in answer to these questions.

1 Why do you want this job?
2 What skills will you bring to this job?
3 Why do you want to leave the job you already have?
4 What questions would you like to ask the interviewers?

12 Work in small groups. Pass your CV and your job details around your group. Role-play a job interview. Take turns to be interviewed by the rest of the group.

Review Unit B

08.35	computer activates abort engine
08.36	abort engine ejects crew capsule from rocket
08.37	abort engine burns out
08.38	attitude-control engine activates
08.45	explosive bolts detonate
08.47	jettison engine fires
08.49	crew capsule separates from LAS
08.53	parachutes open

1 Practise this dialogue with a partner. Refer to the checklist of times and events on the left.

A and B are two engineers working at a Space Centre on Earth. They are monitoring the operation of the Launch Abort System of a rocket soon after its launch.

A: It's **08.36**. What's happened?
B: The computer has activated the abort engine.
A: What time did this happen?
B: Let's see ... the computer activated the abort engine at 08.35.
A: What's happening now?
B: The abort engine is ejecting the crew capsule from the rocket.
A: What'll happen next?
B: The abort engine will burn out.

2 Complete these dialogues in the same way, referring to the checklist above.

A: It's **08.38**. What's happened?
B: The abort engine has (1) __________.
A: What time did this happen?
B: Let's see ... It (2) __________ at 08.37.
A: What's happening now?
B: The attitude-control engine (3) __________.
A: What'll happen next?
B: The explosive bolts (4) __________.

A: It's **08.47**. What's happened?
B: The explosive bolts (5) __________.
A: What time did this happen?
B: Let's see ... They (6) __________ at 08.45.
A: What's happening now?
B: The jettison engine (7) __________.
A: What'll happen next?
B: The crew capsule (8) __________ from the LAS.

3 Work in pairs. Practise a similar dialogue which begins as follows.

A: It's **08.49**. What's happened?
B: The jettison engine ...

4 Complete these news items. Use the present perfect in the *news in brief* (B) and the past simple in the *news in detail* (D).

1 B An earthquake __________ (destroy) part of a town in Italy.
D The earthquake, 6.5 on the Richter scale, __________ (strike) the town in the middle of last night.

2 B Hurricane Rita __________ (hit) the coast of Mexico.
D The hurricane __________ (reached) the first coastal town at around 10.40 yesterday evening.

3 B The first plastic plane __________ (take off) safely.
D The first aircraft with a plastic fuselage __________ (take off) from Frankfurt airport at 7.30 this morning.

5 Rewrite these sentences to give the same or similar meaning, using the word(s) in brackets.

Example: *1 After sounding the fire alarm, the computer closes all the fire doors.*

1 First the central computer sounds the fire alarm. Then it closes all the fire doors. (After / sounding)
2 The explosive bolts detonate, and then they break apart. (As soon as / have)
3 Once the drogue parachute has opened, it pulls the main parachute out of the pilot's backpack. (opening)
4 When the fuel in the cylinder ignites, the burning gas expands rapidly. (As soon as / has)
5 When the drill bit passes 350 metres, it cuts into oil-bearing rock. (Once / has)
6 The sensor detects a temperature change. Then it sends the data to the thermostat. (detecting)

6 Work in pairs. Discuss this situation and the four solutions. Then explain to the class what you and your partner would do if you were in the same situation.

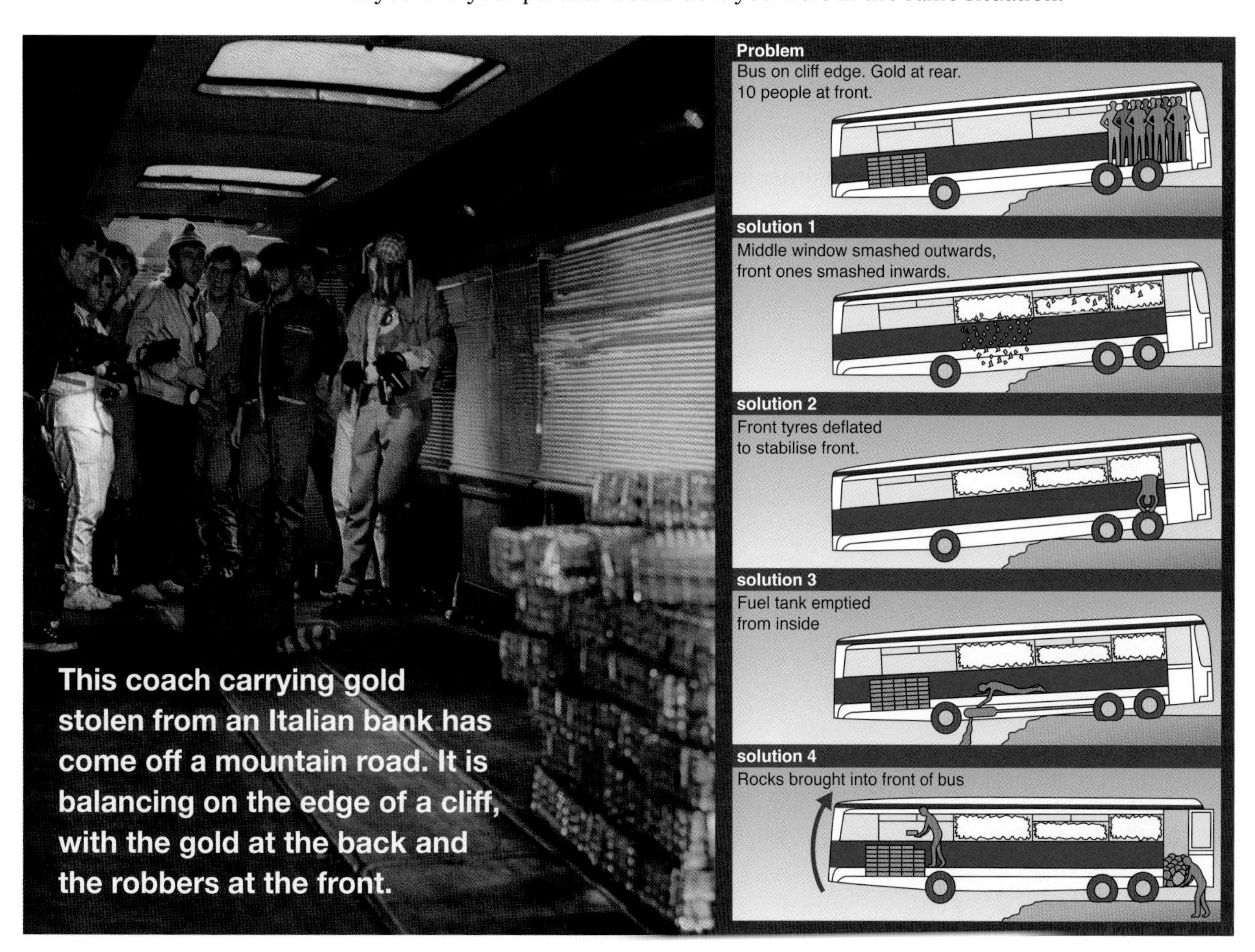

7 Discuss some of these questions with your partner. Then tell the class about your ideas. Use the second conditional where appropriate.

What would happen, or what would you do ...

1 if you could go back in time, or forward in the future?
2 if our universe was only one of many parallel universes?
3 if everyone had a personal jet pack for flying around?
4 if someone invented a computer that could read your thoughts?
5 if CCTV cameras and microphones monitored everything you say and do?
6 if you went to another planet to learn advanced skills, and then returned to Earth?

8 Write the correct noun from the verb in brackets to fill each gap.

1 If the ______ of the parachute fails, the pilot can use the backup. (deploy)
2 The small wing at the rear of the aircraft provides ______ (stabilise).
3 The aircraft will begin its ______ (descend) in five minutes.
4 The engine on the stern of the boat provides massive ______ (propel).
5 The ______ (detonate) of the explosive bolts releases the canopy.
6 Car seat belts act as a powerful ______ (restrain) for the body in a collision.

9 Complete this dialogue.

A: What (1) ______ (you / do) for a living?
B: I (2) ______ (work) for myself. I'm an inventor. I (3) ______ (modify) existing products to improve them.
A: So what product (4) ______ (you / modify) at the moment?
B: Well, right now I (5) ______ (work) on a mobile phone. I (6) ______ (try) to make it smaller, and to give it longer battery time.
A: That's interesting. (7) ______ (you / go) to produce a prototype?
B: Yes. I (8) ______ (present) one to my clients early next year.
A: That's great. Good luck with your presentation.

10 Read the information on page 112 and then write out a list of questions about the structure.

Ask about *shape*, *dimensions*, *number of containers*, *purpose or function*, *method of operation*, *amount of energy used*, *reason for low energy consumption* and any other matters of interest.

11 Work in pairs. Practise asking and answering questions about the structure.

12 Match the headings 1–8 on this form with the correct information a–h.

1 Full name	a) professional footballer
2 Present employer	b) more than 330 games and more than 130 goals.
3 Responsibilities	c) Cristiano Ronaldo dos Santos Aveiro
4 Experience	d) after retiring from playing, to become an international manager
5 Qualifications	e) to help his team to win; to score goals
6 Competences	f) Portuguese
7 Career plans	g) goal-scoring ability; leadership skills
8 Nationality	h) Real Madrid Football Club

13 Write a paragraph comparing these two 2010 World Cup stadiums.

Durban
(top photo)

Capacity* during World Cup	70,000
Capacity after World Cup	58,000
Height of roof (incl. arch)	151 m**
Total length of stadium	320 m
Total width of stadium	280 m

Cape Town
(bottom photo)

Capacity during World Cup	68,000
Capacity after World Cup	55,000
Height of roof	50 m
Total length of stadium	290 m
Total width of stadium	265 m

**capacity* = maximum number of spectators
**there is a 106-m high arch above the roof

14 Complete this dialogue.

A: Where (1) ________________ (you / work)?
B: I work at AutoWorld. I'm a technician there.
A: How long (2) ________________ (you / work) there?
B: (3) ________________ (I / be) there for four years now.
A: That's a long time. Where (4) ________________ (you / work) before that?
B: I (5) ________ (be) a junior technician at MultiPart. I enjoyed it very much.
A: So why (6) ________________ (you / leave)?
B; Because I (7) ________________ (want) to work in a bigger company.
A: By the way, (8) ________________ (you / finish) your technician's diploma?
B: Yes, I have.
A: When (9) ________________ (you / complete) it?
B: Three months ago.
A: Well done.

Project **15** Research a job and an employer you are interested in.

- Search for a suitable job advert.
- Research the company which placed the advert.
- Think of questions the interviewers may ask, and prepare answers.
- Prepare a list of the questions you would like to ask.
- Write a letter of application to accompany your CV.
- Put everything into a special folder. Keep it for future reference.

5 Safety

1 Warnings

Start here 1 Brainstorm with the whole class. What car safety systems have you heard about? Which ones (a) take control from the driver and (b) only warn the driver but do not take control?

Reading 2 Read this article and answer the questions below.

LANE KEEPING ASSIST

a warning system that helps drivers stay in lane

Lane Keeping Assist (LKA) systems help a driver to keep their vehicle in its lane. There are two main types:
- systems which give the driver an audible or visible warning if the vehicle drifts out of its lane. 5 Alternatively, they give feedback such as vibrating the steering wheel.
- systems which also take remedial action after giving a warning. For instance, if the driver ignores the warning, the system automatically corrects the 10 steering so that the vehicle maintains its position in the lane.

Some cars made by Volvo, BMW, GM and Mercedes-Benz, for example, use the warning only system. Using cameras that monitor road markings, the system can 15 detect when a car departs from its lane.

The system can also decide if the driver leaves the lane intentionally. It gives no warning if the driver accelerates before overtaking, brakes heavily, activates the indicators, steers into a bend, or returns the car to its lane.

20 However, if the system decides that the driver is moving the car out of its lane unintentionally, it activates a motor, which vibrates the steering wheel. This gives gentle tactile feedback, and usually makes the driver counter-steer until the car regains its position in the lane.

25 Some Toyota LKA systems, however, intervene if the driver ignores a warning. For example, it may apply some pressure on the brakes or some torque on the steering column to bring the vehicle back into the lane.

1. Can you think of examples of (a) *audible* (line 3), (b) *visible* (4) and (c) *tactile* (23) warnings?
2. Which make of car takes corrective action if the driver does not respond to a warning?
3. If a driver uses the car's indicators and then crosses into another lane, what action, if any, does the warning system take?
4. Find words in the text that mean (a) *corrective* (b) *check continuously* (c) *discover* (d) *on purpose* (e) *steer in the opposite direction* (f) *twisting motion*

Vocabulary 3 Find six more phrases in the text which express the meanings in the table (two for each meaning).

remain in correct lane	leave the lane	go back into lane
stay in lane	*drifts out of its lane*	*returns the car to its lane*

Listening **4** [14] Listen to this phone conversation and answer the questions.

1 What is Max's *main* purpose for phoning Tom?
2 What topics do they talk about *before* and *after* discussing the main purpose of the call?
3 After this call, do you think Max and Tom's working relationship is (a) worse (b) the same or (c) better than before? Why do you think this?

5 [15] Listen to this meeting and answer the questions.

1 What is the purpose of the meeting?
2 Which (a) sensor and (b) warning system do they agree on?

6 Listen again and answer these questions.

1 What type of sensor was suggested but not agreed on?
2 What *four* types of warning were suggested but not agreed on?
3 Why does Tom not like safety systems which take control of the car?

7 Complete these extracts from the discussion with the words and phrases in the box.

by the way | in other words | alternatively | you have a point | anyway | for instance

1 Yes, I agree, ________________ there …
2 … the controller can give a warning. ________________, you could have a flashing light …
3 We shouldn't use a SatNav voice. ________________, I had a very bad experience with a SatNav last week …
4 ________________, let's keep to our main discussion.
5 … we should use normal feedback signals, ________________, signals from the real world.
6 It could make him … press the brakes too hard. ________________, he might counter-steer too much …

8 Listen again, and check your answers to 7.

9 Match the language functions 1–6 with the words and phrases in 7.

1 give an example
2 change the subject
3 say the same thing differently
4 agree with someone
5 give another possibility
6 return to the main point

Speaking **10** Work with the whole class. Discuss the question 'Are cars too safe?' The class should divide into two groups with opposite points of view.

Group A: Your job is to design automatic safety devices (such as anti-lock braking devices and air bags). Your view is that everything possible should be done to protect drivers, passengers and other road users by developing automatic safety devices in cars.

Group B: Your job is to advise car manufacturers on driver behaviour in safety matters (similar to Tom's job in audio 14 and 15). Your view is that too many automatic safety devices in a car take away the driver's responsibility. They make the driver feel too safe and protected, which is dangerous. You prefer feedback systems which warn the driver about dangers.

2 Instructions

Start here

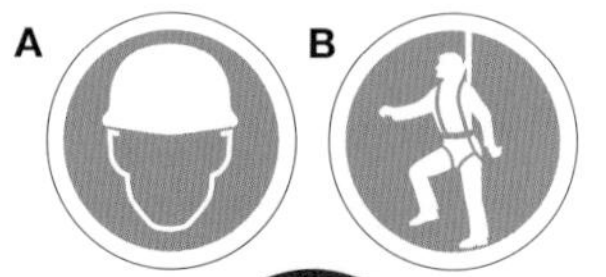

1 Match the signs with the instructions. Which ones show that the action is (a) mandatory or (b) prohibited?

Example: *1–E*

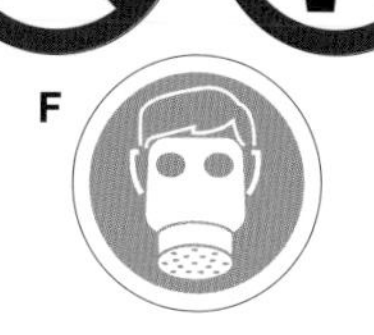

1 Do not drink this fluid.
2 Wear breathing equipment in this area.
3 Wear a safety harness here.
4 Do not extinguish fires here with water.
5 Do not oil or clean this machine when it is moving.
6 Wear a hard hat (safety helmet) on this site.

Listening

2 Work in pairs. Match the labels on this diagram with the parts in the box below.

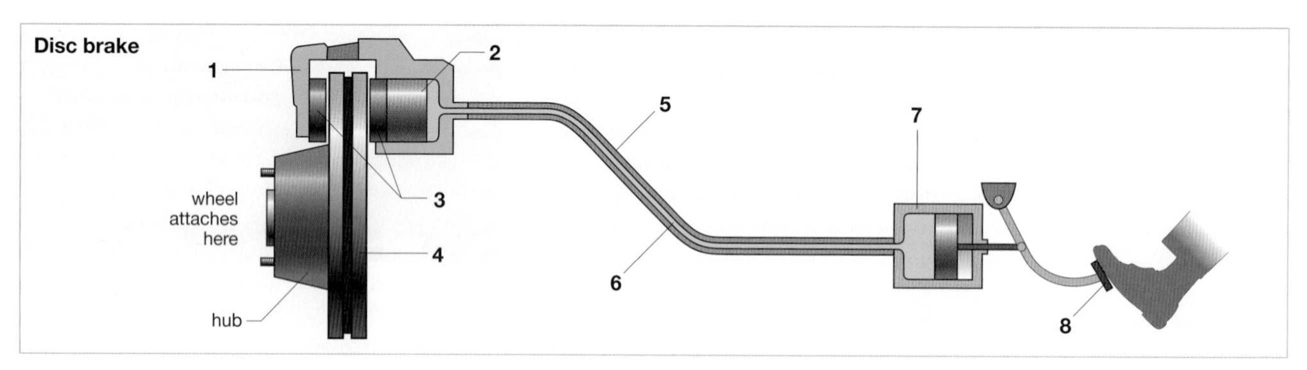

disc piston brake line master cylinder
brake pedal hydraulic fluid brake pads calliper 16

3 17 Listen to this discussion and check your answers to 2.

4 Listen again and answer these questions.

1 What two things does the mechanic say the driver should do?
2 What single thing does the mechanic say the driver must do?
3 What does he say are three possible causes of a soft or spongy brake pedal?
4 Why are the brake pads made of a softer material than the disc?

Language

Modals and semi-modals followed by *active* verbs:					
You	must / should	wear your hard hats.	You	mustn't / shouldn't	take them off yet.
You	have to / need to	silence your mobiles.	You	don't have to / need to	switch them off.

Modals and semi-modals followed by *passive* verbs:					
Helmets	must / should	be worn.	Helmets	mustn't / shouldn't	be taken off yet.
Mobiles	have to / need to	be silenced.	Mobiles	don't have to / need to	be switched off.

5 Make instructions for the safety signs in 1. Use *must, must not, have to* or *need to*. Use the passive where appropriate.

Example: *A Hard hats must be worn on this site.*

6 Make some recommendations for improvements to your college or workplace.

Example: *MP3 players should be given to all students to help with language practice.*

7 Describe some things which are unnecessary in your college or workplace, or in your life in general.

Example: *I don't have to get up early on Tuesdays because classes start at 10.00.*

Reading **8** Read the manual and say which actions 1–12 are essential (E), recommended (R) or unnecessary (U).

The complete brake system (including discs, callipers, pads, pistons and brake lines) should be inspected at least once a year (1). The fluid level should be topped up and the handbrake should be adjusted at the same time (2).

Brakes should normally be replaced after 20,000–30,000 kilometres (3). Of course, they don't have to be replaced if inspection shows they are in good working order (4).

However, brakes must not be used when the brake pads are below the minimum thickness (5).

Brake lines should be drained, and the brake fluid replaced, at least every two years (6).

In addition to the above, always monitor your braking, and notice anything unusual in the brakes while you are driving.

For example, if your brake pedal feels soft or spongy when you press it, it is possible that air has entered the brake lines. If that is the case, the brake lines need to be drained and have to be refilled with new brake fluid (7).

If the brakes make a loud grinding sound, this probably means that the brake pads are very worn. When this happens, the car must not be driven any further (8). The pads must be replaced, and the discs have to be inspected for damage (9). If there is damage, the discs need to be replaced or mended (10).

On the other hand, if your brakes give a light squealing noise, this may not be serious. There may be water or dust on the brake pads and discs. If so, the pads probably do not need to be replaced (11). However, they should still be inspected (12).

9 Are these statements *true* (T) or *false* (F)? Correct the false ones.

1 The driver should empty out all the fluid from the pipes in the brake system and fill it up with new fluid at least every two years. (T / F)
2 A feeling of softness when you press the brake pedal may indicate that some air has escaped from the pipes of the brake system. (T / F)
3 If you hear a loud noise like metal scraping on metal, it is likely that the brake pads have worn too thin. (T / F)
4 Any noise when you press the brake pedal means that there is a serious problem in the braking system. (T / F)

Vocabulary **10** Match the maintenance verbs 1–8 from the text with their meanings a–h.

1 refill	a) modify to the correct condition
2 inspect	b) empty out
3 top up	c) add more fluid to the correct level
4 replace	d) fill up with new fluid
5 drain	e) check frequently
6 monitor	f) repair
7 mend	g) exchange with a new one
8 adjust	h) examine

Language **11** Rewrite the odd-numbered instructions (1, 3, 5, etc.) in the manual in 8 using the active form. Begin each sentence *You* … . You can leave out some information to shorten the sentences.

Example: *1 You should inspect the complete brake system at least once a year.*

Writing **12** Write a set of instructions for maintaining a machine or device which you are familiar with.

- Include items which are (a) essential (b) recommended and (c) unnecessary.
- Use active verbs where you think the user can do the maintenance themselves.
- Use passive verbs where you think the user should get an expert to do it.

3 Rules

Start here

1 Discuss the situation in the photo and the table with a partner. Can you devise a rule which would prevent it from happening?

	Aircraft A	Aircraft B
Altitude	29,000 feet (FL 290)	29,000 feet (FL 290)
Heading	East (90°)	South East (135°)

Reading

2 Read the rules below. Which one(s) solve the problem in 1?

Rules of the Air (Part 1): Altitude for Direction-of-Flight

The following rules apply when an aircraft is cruising (that is, flying without changing altitude):

1 Aircraft must fly at least 10 flight levels (FLs)* (that is, 1,000 feet) above or below other aircraft.
2 On headings of 360° to 179° (e.g. N, NE, E, SE), aircraft must fly at odd numbered flight levels. For example, aircraft heading east may operate at FLs 170, 190, etc.
3 On headings of 180° to 359° (e.g. S, SW, W, NW), aircraft must fly at even numbered FLs. For example, aircraft heading west may operate at FLs of 160, 180, etc.

* 1 flight level (FL) = 100 feet (30.48 m)

3 A trainer is giving these instructions to a trainee pilot. Tick the ones which follow the rules in 2.

1 Only fly at flight level 180 if your heading is between 180° and 359°. ☐
2 Don't change your heading from NW to NE without changing your flight level from 210 to 220. ☐
3 Only change your heading to E after checking that you are on FL 150. ☐
4 Don't cruise at any flight level before ensuring that you are at least 1000 feet above all other aircraft. ☐
5 Only change your heading from SE to SW when you've climbed from FL 200 to FL 210 ☐
6 Don't cruise at flight level 190 unless you've checked that you're on a heading between 360° and 179°. ☐

Language

+	Only	change your heading	if / when / after	you've checked your flight level.
			after	checking your flight level.

-	Don't	change your heading	unless / until	you've checked your flight level.
			without / before	checking your flight level.

4 Rewrite these instructions to give the same or similar meaning, using the words in brackets. Do not use the words in italics.

1 Your aircraft must *not* turn *unless* it has enough space to do so. (only / if)

Example: *Your aircraft must only turn if it has enough space to do so.*

2 You *have to* turn your aircraft *in a way which does not* endanger other aircraft. (must / without)
3 The slower helicopter must *not* change direction *until* the faster one has passed it. (only / after)
4 A car driver must *only* overtake another vehicle *after* checking his mirror. (not / before)
5 Drivers should *only* start a long car journey *if* they have checked brakes, tyres and fluid levels. (not / unless)

Scanning

5 Practise your speed reading. Look for the information you need on the SPEED SEARCH pages (116–117). Try to be first to complete these sentences.

1 The planes in the near-miss incident were ______ metres apart at their closest point.
2 The incident took place close to __________ International Airport.
3 The pilot of the light aircraft was __________ (nationality).
4 To avoid a collision, the pilot of the Boeing 737 first turned the plane __________ (left / right), then descended to ______ feet, climbed, and finally descended again.

Speaking

6 Work in small groups. Discuss the dangerous situations below, read the rules of the air, and decide (a) which aircraft must take evasive action, and (b) what action it/they must take.

Look again at Rules of the Air (Part 1) on the previous page. Then divide your group into two sub-groups to study Parts 2 and 3. When you have finished, come together as a group. Explain your rules to the other group, then discuss the situations and make your decisions.

Sub-group A: Study Rules of the Air (Part 2) on page 110.
Sub-group B: Study Rules of the Air (Part 3) on page 115.

Situation 1

	Aircraft A	Aircraft B
Altitude	FL 800	FL 900
Heading	NE (45°)	SW (225°)
Phase	Climbing	Descending

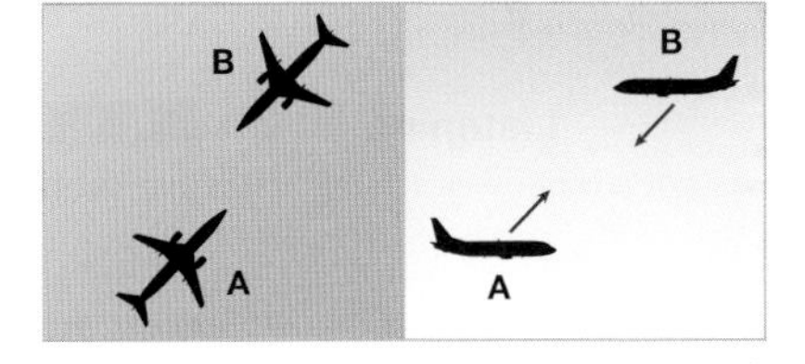

Situation 2

	Aircraft A	Aircraft B
Altitude	FL 170	FL 170
Heading	E (90°)	NW (315°)
Phase	Cruising	Cruising

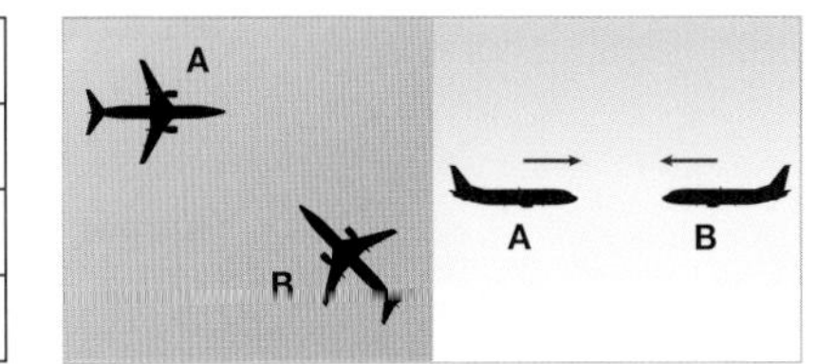

7 Present your group's decisions to the rest of the class and give reasons for them.

Writing

8 Write the most important rules for an activity, procedure, sport or game that interests you.

6 Planning

1 Schedules

Start here 1 Brainstorm with the whole class. Think of as many sources of energy as possible including the ones in the photos. Which ones are (a) renewable, (b) non-renewable, (c) carbon-based or (d) non-carbon-based. Make notes.

Listening 2 18 Listen to the first part of this meeting and tick the greenhouse gases which are mentioned.

ozone (O_3) ☐ nitrous oxide (N_2O) ☐
methane (CH_4) ☐ carbon dioxide (CO_2) ☐

3 19 Listen to the rest of the meeting, and mark the deadline (final date) for each target on the Gantt chart. You can shade in the chart later.

Target	Year: 2014, 2015, and so on							
	14	15	16	17	18	19	20	50
1 Cut CO_2 emissions by 50%								
2 Bring down CO_2 emissions by 20%								
3 Reduce overall energy consumption by 20%								
4 Switch 30% of coal-fired power to CCS								
5 Convert 10% of transport fleet to bio fuel								
6 Meet 20% of energy needs from renewables								

Note: *CCS* = carbon capture and storage

4 Complete these phrases used in the meeting in 3. Listen again and check your answers. Which ones mean (a) I agree and (b) I disagree?

1 OK, I can go ______ ______ that.
2 I would ______ with you there.
3 You have a good ______ there.
4 I'm not ______ about that deadline.
5 That ______ about right.
6 You're absolutely ______.
7 I'm in complete ______ with you.
8 I can't ______ ______ with that.

5 What other phrases to express agreement and disagreement can you remember from the meeting? Look at the audio script on pages 121–122 to help you.

Speaking

6 Work in groups. Discuss the energy sources you noted in 1. Exchange any strong opinions you may have about the use of these sources in the future.

Vocabulary

Here are some ways to set a *deadline* (the last possible date for doing a task):

- Please finish the report *by* the end of June (at the latest).
- We have to start the project *no later than* 15th March.
- *The deadline* for completing the investigation is the end of the year.

7 Rewrite these statements using the words in brackets.

1 We'll have to finish this job no later than June 10th. (by)
2 You have to complete the project by the end of the month. (The deadline)
3 Please finalise the report by next Monday. (The last possible date / is)
4 The deadline for switching to bio fuels is next year. (will have to / no later than)

Language

Verbs like *can*, *must*, *have to*, *need to* normally refer to both present and future time. However, if we want to emphasise that they refer to the future, we can use the forms in the right-hand column

now or in the future	emphasis on the future
*We **can** do it.*	*We**'ll be able to** do it (by 2019).*
*They **can't** do it.*	*They **won't be able to** do it.*
*I **must** / **have to** / **need to do it**.*	***I'll have to** / **need to** do it.*
*You **don't have to** / **need to** do it.*	*You **won't have to** / **need to** do it.*

8 Rewrite these statements to emphasise the future. Replace the words in italics with phrases using words in the box.

will won't going to be able to have to need to

Example: *1 The world is going to have to (or will need to) make a 40% reduction in emissions before 2040.*

1 The world *must* make a 40% reduction in emissions before 2040.
2 Our company *can't* meet the 20% target unless we convert to carbon capture.
3 We *can* probably convert half of our power plants in the next ten years.
4 The 2020 deadline is too tight. I'm sure we *can't* meet it.
5 Companies *mustn't* continue consuming the same amount of energy in the future.
6 Our company *doesn't have to* cut emissions by 50%. The target is 20%.
7 In future we *must* obtain at least 50% of our energy from renewable sources.
8 People *don't need to* stop flying completely, but they *must* do it less.

Speaking

9 Draw a Gantt chart to show the main stages and deadlines in a project which is important in your own work, studies, career or life in general.

10 Explain your project to the rest of the class or to a group.

2 Causes

Start here 1 Work in small groups. Answer the questions about the four illustrated processes.

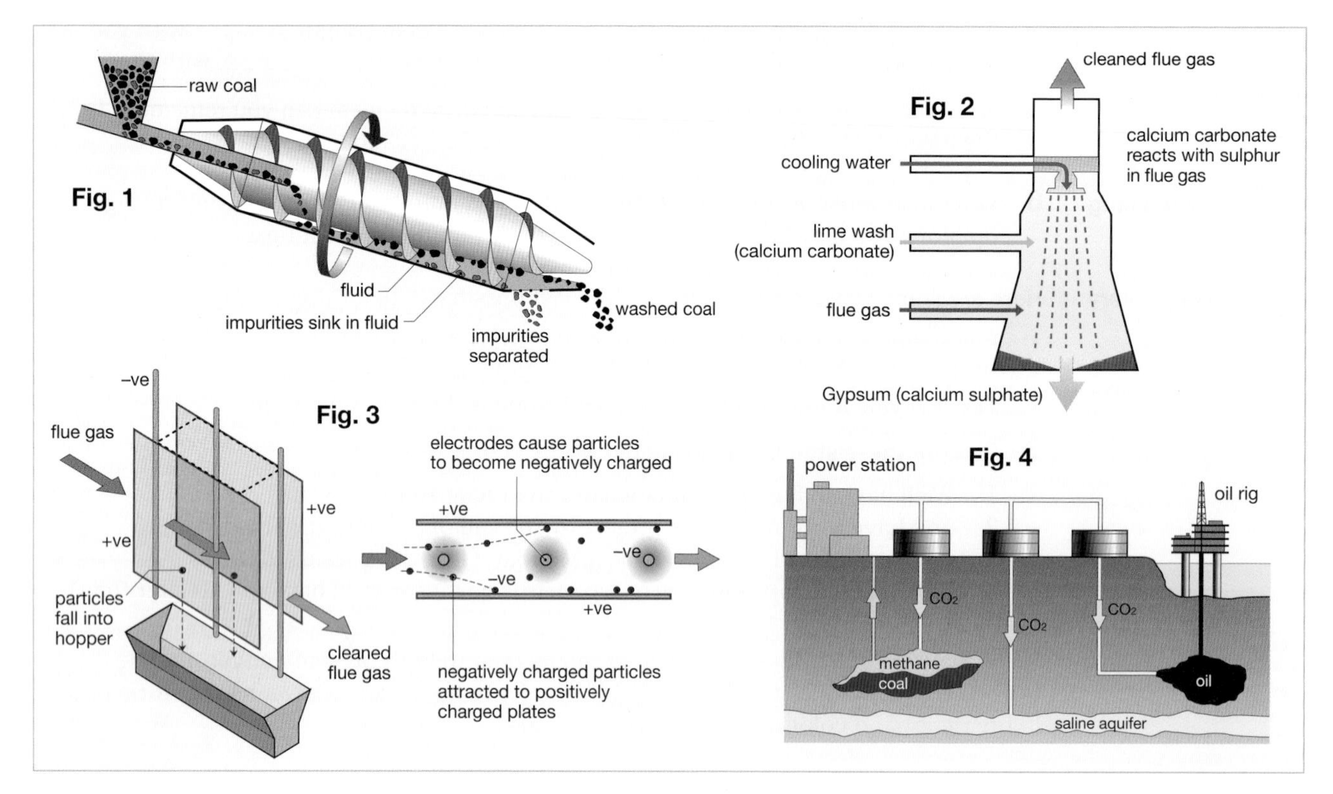

1 What is the main purpose of the four processes? (The answer concerns *coal*.)
2 Look at Fig. 4. What *extra* benefits does this process bring? (The answer concerns *oil* and *methane*.)

Scanning 2 Practise your speed reading. Look for the information you need on the SPEED SEARCH pages (116–117). Try to be first to complete these statements.

1 There are approximately ____________ tonnes of coal in the world.
2 The coal in the world will last for a maximum of ____________ years.
3 Particulates are removed from flue gas by electrostatic ____________.

Reading 3 Match the questions in 1–6 with the answers in a–f, and with the diagrams in 1.

1 How are particles removed from flue gas?
2 How is sulphur separated from flue gas and removed from it?
3 How do you take impurities out of small pieces of coal?
4 Isn't it dangerous to store CO_2 in underground saline aquifers?
5 How does carbon capture help us to recover methane from underground coal fields?
6 Why do you say that CCS can help the oil industry to get oil from old oil wells?

a) The *storage* of CO_2 in the aquifer is harmless owing to the *presence* of salt in the water, and because of the depth of the aquifer.
b) The *force* of the oil *rise* is due to the *pressure* from the CO_2.
c) The rise of methane to the surface and its *recovery* from the ground are due to the pressure of the CO_2 in the underground coalfield.
d) The *desulphurisation* of the flue gas happens as a result of a chemical *reaction*. This is due to the *insertion* of calcium carbonate and the *addition* of water to the gas.
e) The *removal* of polluting particles from the flue gas is caused by their *attraction* to the collection plates. This is due to the negative electric *charge* that they get from the electrodes.
f) The *purification* of the pulverised coal is due to the *rotation* of the barrel and the density of the fluid.

Vocabulary

5 Change the nouns in *italics* in a–f in 3 into verbs. (Sometimes the verb and the noun have the same form).

Example: *a) store; be present*

6 Study the information in the table. Then guess the meanings of the words below. Say whether they are nouns or verbs. When you have finished, check your answers in a dictionary or reference book.

These suffixes indicate *causation*.	
verb	**noun**
-ify *purify* (= to make something pure)	*-ification* *purification* (= the process of purifying)
-efy *liquefy* (= to change something into liquid)	*-efaction* *liquefaction* (= the process of liquefying)
-ise (BrE); *-ize* (AmE) *sulphurise* (= to make something contain sulphur)	*-isation* (BrE); *-ization* (AmE) *sulphurisation* (= the process of sulphurising)

Note: BrE *sulphur sulphurise sulphurisation*; AmE *sulfur sulfurize sulfurization*

1 *humidify* (noun / verb) ______
2 *ionisation* (noun / verb) ______
3 *ozonification* (noun / verb) ______
4 *gasification* (noun / verb) ______
5 *solidify* (noun / verb) ______
6 *pulverise* (noun / verb) (Note: *pulver-* is Latin for *dust*)
______

Language Ways of expressing causation

using a verb		using a noun	
because	*the barrel* ***rotates*** *water* ***is added*** *the chemicals* ***react*** *the CO_2* ***pressurises*** *the oil*	due to owing to caused by as a result of	*the* ***rotation*** *of the barrel* *the* ***addition*** *of water* *the* ***reaction*** *of the chemicals* *the* ***pressure*** *of the CO_2 on the oil*

7 Rewrite each sentence to give a similar meaning, making these changes

- replace *because* with the phrase in brackets.
  Example: *because* ➔ *owing to*
- replace the verbs in italics with related nouns.
  Example: *rotate* ➔ *rotation*

Example: *1 We have to use international time zones owing to the rotation of the earth.*

1 We have to use international time zones because the earth *rotates*. (owing to)
2 There's no need to pump the oil, because the CO_2 *pressurises* it. (due to)
3 The iron filings are moving because the magnet *is attracting* them. (as a result of)
4 The pressure on the methane is because CO_2 *is injected*. (caused by)
5 People must not drink this water because impurities *are present*. (owing to)
6 This concrete has flaws because too much water *was added*. (caused by)
7 Our astronauts are safe because the capsule *was recovered* from the sea. (due to)
8 We emit no carbon because our CO_2 *is stored* underground. (as a result of)

3 Systems

Start here 1 Work in small groups for all the speaking exercises in this section. Briefly discuss the illustration with the rest of the group.

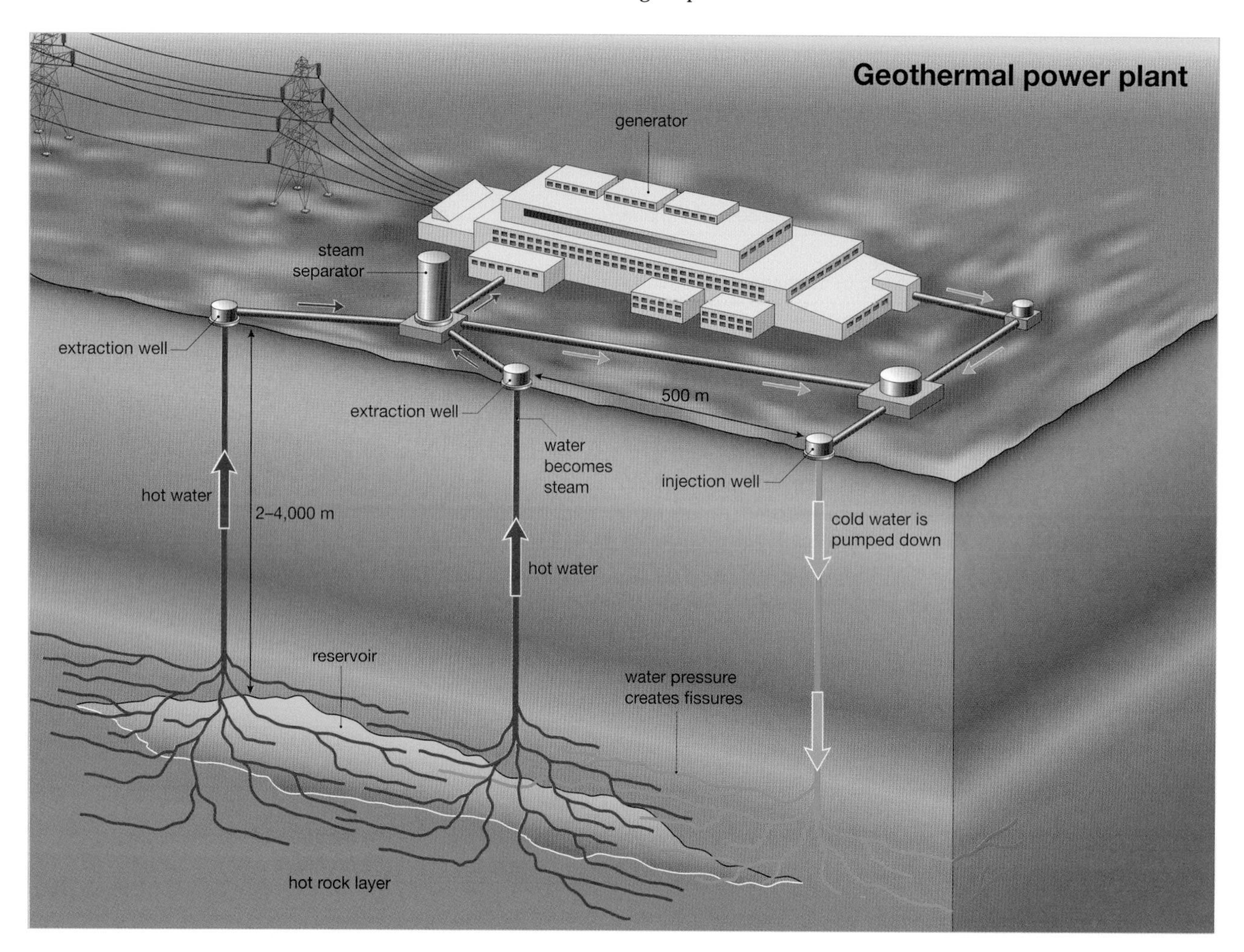

Speaking 2 You are going to prepare a short talk with your group about how the system works. As a first step, discuss the answers to these questions, and make notes.

1. Where is the power plant located?
2. What are the four main elements of the power plant?
3. What is the minimum distance between the injection and extraction wells?
4. Why is there this minimum distance between the injection and extraction wells?
5. How does the cold water get down to the hot rock layer below the surface?
6. What causes the fissures to form below the surface?
7. How is the reservoir of hot water formed?
8. What forces the hot water to the surface? (temperature? pressure?)
9. What equipment enables the hot water to rise to the surface?
10. Where does the hot water change into steam?
11. What causes the hot water to change into steam? (pressure)
12. What does the steam separator do?
13. What happens to the steam on the surface?
14. After the steam has condensed into water, what happens to the water ?

3 Divide your notes into sections.

For example, one section of the talk could be about the *injection stage* of the process, in which water is injected down from the surface to the layer of hot rock.

4 Complete these phrases with the words in the box. Then add them to your notes at the appropriate points.

let's look at	let's move	like to move on to	I'm going to
let's look	aim in this talk	I'd like to thank	I'd like

1. My __________ is to explain how geothermal energy works.
2. First of all, __________ to discuss the location of a geothermal power plant.
3. OK. Now __________ on to the main elements of the power plant.
4. Right, so now __________ talk you through the whole process.
5. And to begin with, __________ at the injection stage of the process.
6. All right, now I'd __________ the extraction stage of the process.
7. So finally, __________ the surface stage of the process.
8. Well, __________ you all for listening. Any questions?

5 Prepare the talk with your group.

Divide up the tasks between all the group members. One way of doing this would be to have a different person to present each section of the talk, and then the rest of the group answers questions from the class.

6 Give your talk (as a group) to the class, and answer any questions.

When you hand over to the next person in your group, use a phrase such as this: *Now I'd like to ... I'm going to hand over to my colleague to talk about ... to answer your questions.*

Writing

7 Work individually. Write a reply to this email.

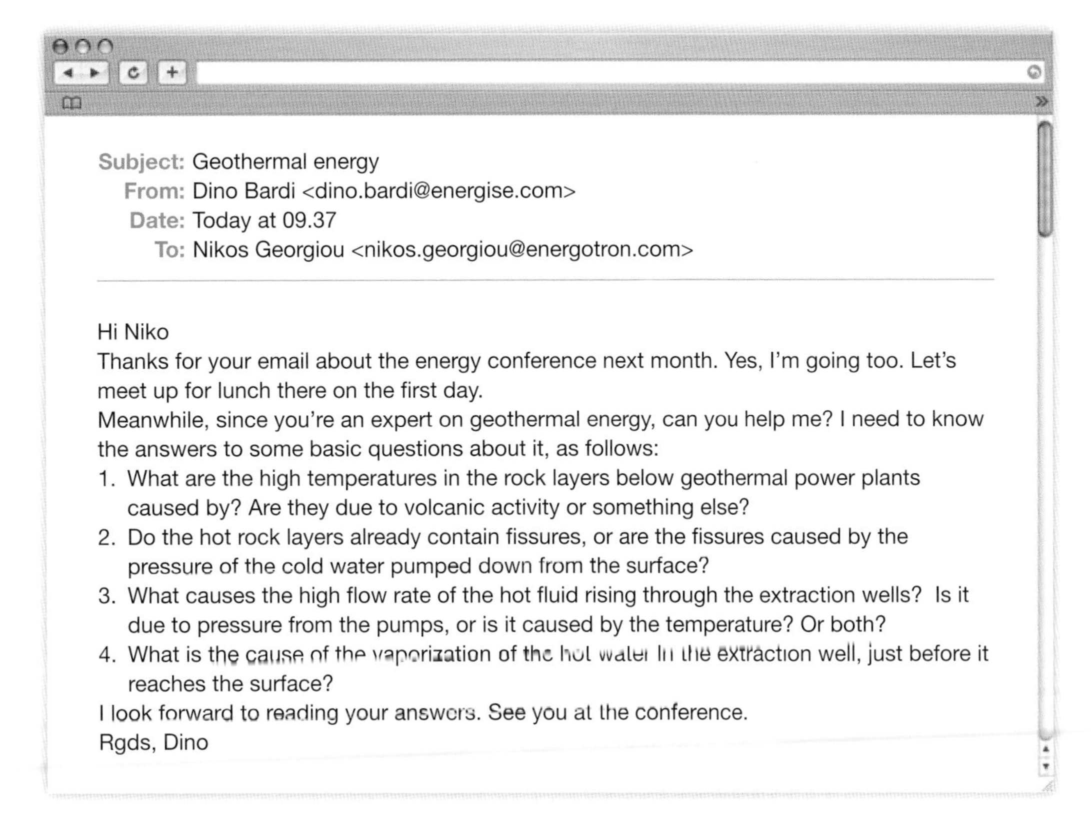

Subject: Geothermal energy
From: Dino Bardi <dino.bardi@energise.com>
Date: Today at 09.37
To: Nikos Georgiou <nikos.georgiou@energotron.com>

Hi Niko

Thanks for your email about the energy conference next month. Yes, I'm going too. Let's meet up for lunch there on the first day.

Meanwhile, since you're an expert on geothermal energy, can you help me? I need to know the answers to some basic questions about it, as follows:

1. What are the high temperatures in the rock layers below geothermal power plants caused by? Are they due to volcanic activity or something else?
2. Do the hot rock layers already contain fissures, or are the fissures caused by the pressure of the cold water pumped down from the surface?
3. What causes the high flow rate of the hot fluid rising through the extraction wells? Is it due to pressure from the pumps, or is it caused by the temperature? Or both?
4. What is the cause of the vaporization of the hot water in the extraction well, just before it reaches the surface?

I look forward to reading your answers. See you at the conference.

Rgds, Dino

8 Exchange emails with a partner. Check your partner's email for grammar, spelling and punctuation.

Review Unit C

1 Work in pairs. Discuss these questions. Then write the answers.

1. What has happened? What is happening now? What is going to happen?
2. What will happen if the larger boat (a) turns left (b) turns right?
3. What action would the smaller boat probably take if it had an engine?
4. What rule(s) should the crews of the boats follow to avoid this situation?

2 Match the phrases in italics in the dialogue with phrases in the box with the same or similar meanings.

by the way	in other words	alternatively	you have a point there	anyway	for instance

A: I think that cars need more automatic safety devices. There are already too many accidents on the road.
B: Well, (1) *I agree with that*, but I don't think that more automatic devices will make drivers safer.
A: Why not? I'm thinking about devices that prevent accidents, (2) *such as* lane keeping assist systems.
B: (3) *To change the subject for a moment*, have you seen the latest accident statistics?
A: Yes, unfortunately they're higher than last year's.
B: Yes, they are. (4) *But to return to the subject*, I prefer systems that give alerts to the driver, (5) *that is*, systems that warn the driver that they're driving dangerously.
B: (6) *Or, to consider a different possibility*, perhaps we need systems that give warnings, and then slam on the brakes if the driver ignores them.

3 Practise the dialogue in pairs, using the phrases from the box in 2.

4 Complete these sentences with the correct form of words in the box. Use each word once only.

remain	keep	maintain	depart	regain

1. The plane ____________ the same speed for the whole of the flight.
2. The temperature in the hotel ____________ at 22°C all day because of the new thermostat.
3. The ship drifted off course, but later ____________ its correct position.
4. The valve ____________ high pressure on the fluid continuously for 35 minutes, and then opened and released the pressure.
5. The train ____________ from the station on time and reached the next station early.

5 Change these sentences from the active to the passive form. Do not use *by* plus the agent. After each sentence, write E (*essential*), R (*recommended*) or U (*unnecessary*) depending on the modal verb used.

Example: *1 Bottles of liquid must not be taken through airport security. (E)*

1 Passengers mustn't take bottles of liquid through airport security. ☐
2 You don't have to wear your seat belts while the aircraft is cruising. ☐
3 People shouldn't eat large quantities of food during the flight. ☐
4 Everyone has to switch off their mobile phones during take-off and landing. ☐
5 People don't need to show their passports when leaving this airport. ☐
6 Passengers should drink plenty of water during a long flight. ☐

6 Write safety instructions which are important in your technical field.

- Write four which you can easily carry out yourself. Use active verb forms.
- Write four which need someone more expert than you to carry out. Use passive verb forms.

7 Write six sentences explaining what has to be done in this checklist. Use *must*, *have to* and *need to* at least once each with passive verb forms.

Example: *1 Before long journeys, the brake fluid level must be inspected and …*

before long journeys	*inspect brake fluid level and top up if necessary*
every 10,000 km	*drain out old engine oil and replace with new oil*
every six months	*inspect brakes and adjust if necessary or replace if worn*
during journeys	*monitor performance of brakes and take any necessary action*
every 15,000 km	*check brake lines and mend or replace if damaged*
before every winter	*drain radiator and refill with mixture of water and anti-freeze*

8 Rewrite these instructions to give the same meaning. Use the words or phrases in brackets.

Example: *1 Food must not be taken on site unless the manager has given permission.*

1 Only take food on site if the manager has given permission. (Food must not …)
2 Only handle those dangerous chemicals if you're wearing protective gloves. (Those dangerous chemicals should not …)
3 Don't walk onto the site before putting on a hard hat. (must / until)
4 You don't need to wear these safety boots unless you're going to work with live electrical equipment. (These safety boots don't have to … / if)
5 Don't start a long car journey before topping up all fluid levels. (should only / after / topping up)

9 Write an explanation of how a room thermostat works.

Begin: *The temperature sensor in the room detects …*

10 Work in pairs. Discuss this question.

In what way(s) is the operation of a room thermostat similar to the operation of the Lane Keeping Assist car safety system described on page 36?

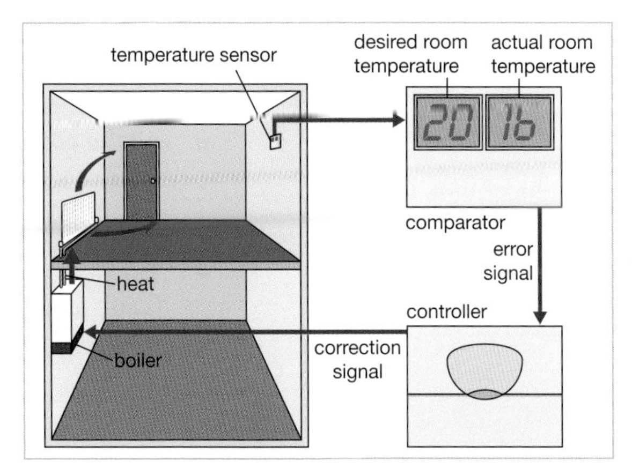

11 Choose the correct words in the box to complete the dialogue.

agree disagree agreement disagreement would should have go am are absolutely sure

Ben: I think we (1) ____________ switch our main energy source from oil to coal.
Sue: I'm in complete (2) ____________ with you. Coal is cheaper to extract than oil. Do you (3) ____________, Ali?
Ali: Yes, you're (4) ____________ correct, it's cheaper. But I'm not (5) ____________ about coal. It's quite a dirty resource. It emits a lot of carbon.
Petra: You (6) ____________ a point there, Ali. But we can capture the carbon and pump it underground. We should invest in CCS (carbon capture and storage).
Jose: I can (7) ____________ along with that. But the technology is still new and untried.
Ben: I (8) ____________ agree with you there. But I still think coal is the future.

12 Make full sentences from these notes. Use all the words in the box at least once. Change the nouns in italics into verbs in the active form.

will won't going to have to need to able to by

Example: *1 The builders will have to construct the foundations by 30th May.*

1 *construction* of foundations by builders – deadline 30 May – ESSENTIAL
2 *removal* of waste by local council workers – deadline 13 June – UNNECESSARY
3 *recovery* of wrecked ship by divers – deadline end tomorrow – IMPOSSIBLE
4 *activation* of new equipment by technicians – deadline next week – ESSENTIAL
5 *transmission* of signal by radio operator – deadline 12.30 today – POSSIBLE
6 *installation* of wiring by electricians – deadline end today – UNNECESSARY
7 *extraction* of ores by mining company – deadline end year – ESSENTIAL
8 *insertion* of new tubes in well by drill team – deadline this week – UNNECESSARY

13 Write the verbs related to these nouns.

1 consumption *consume*
2 emission ____________
3 purification ____________
4 desulphurisation ____________
5 humidification ____________
6 liquefaction ____________
7 solidification ____________
8 gasification ____________

14 Underline the correct words or phrases in brackets to complete the sentences.

1 The electrician was protected from electric shock (owing to / because) his rubber safety boots.
2 A tsunami is a huge tidal wave (causing / caused by) an earthquake below the seabed.
3 The Columbia space shuttle was destroyed (due to / because) a piece of insulating foam fell off and struck the shuttle's wing during the launch.
4 The Titanic sank quickly after striking an iceberg (as a result / because of) its single-hull design.
5 The Air France Concorde caught fire and crashed (caused by / because) it struck a piece of debris on the runway.
6 All the computers in our company have crashed (as a result of / resulting in) a serious virus from an email.
7 The escape of huge quantities of oil from the damaged oil well was (due to / a cause of) a fracture in one of the pipes.
8 The generator at the geothermal plant broke down (because / because of) some faulty wiring in the main circuit.

15 Change the captions below into full sentences about the process in this illustration. Change the nouns in italics into verbs in the passive form.

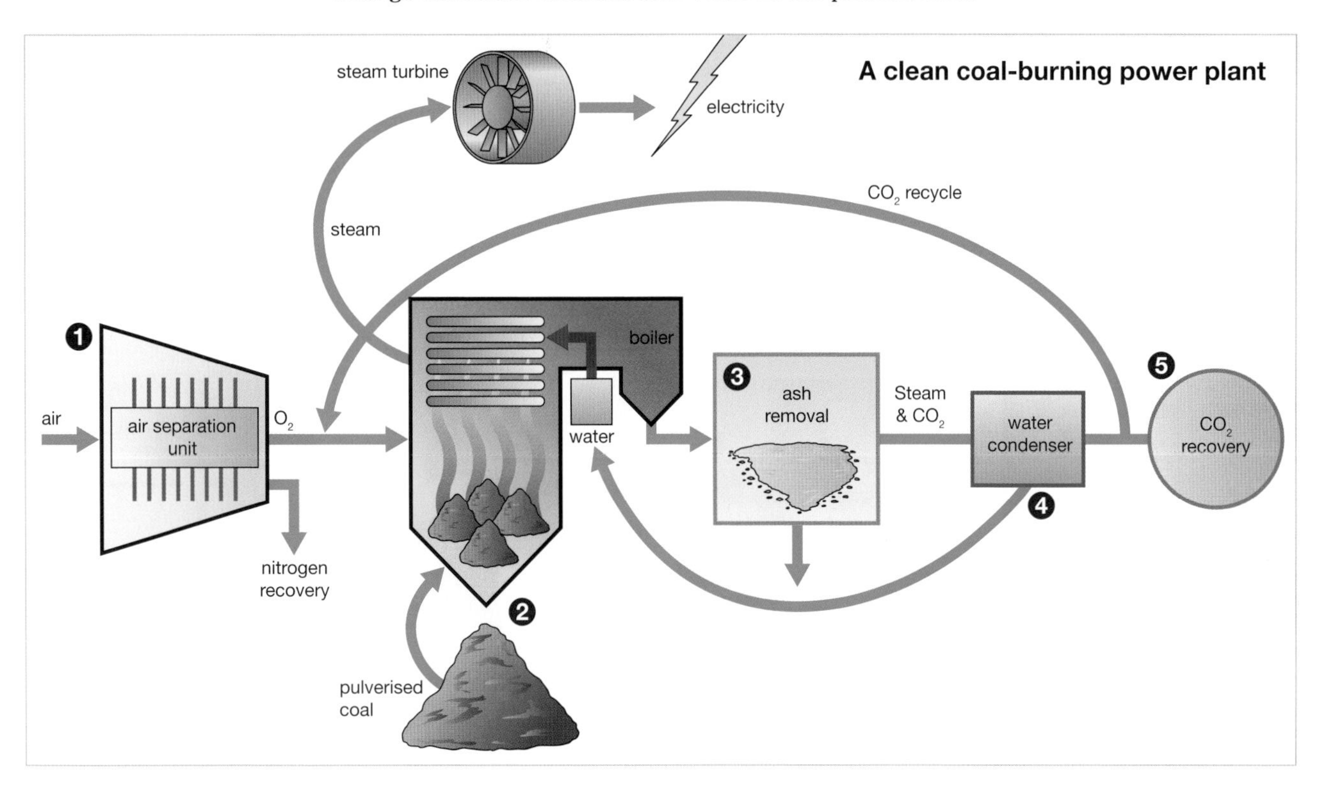

Example: *1 Oxygen is separated from air and introduced into the boiler.*

1. *Separation* of oxygen from air. *Introduction* of oxygen into boiler.
2. *Pulverisation* of coal. *Combustion* in oxygen-rich environment. *Reduction* of pollutants.
3. *Removal* of ash using electrostatic precipitators.
4. *Condensation* of steam into water. *Transfer* of water to boiler. *Completion* of combustion using water.
5. *Recycling* of CO_2 emission from combustion. *Recovery* of CO_2 for later capture and storage.

16 In pairs, practise giving a talk about the coal-burning power plant. Introduce the whole talk, and then each stage of the process, using the information in 15. Refer to the numbers (1–5) in the illustration, which correspond to the five captions in 15.

Here is one way to begin the talk. Use different expressions in your own talk.

The aim of my talk is to describe how a clean coal-burning power plant works. First of all, let's look at the first stage of the process. As you can see in point 1 in the diagram, oxygen is separated from the air, and the oxygen is introduced into the boiler. Now I'd like to move on to the second stage …

17 Write a description of the process you talked about in 16.

Projects **18** Research one of the following:

1. A technology or industry which has had safety issues, or a serious incident involving health and safety. Find out what happened and why it happened. Then work out some rules to prevent similar problems in the future. Write up your findings and recommended safety rules.
2. A national or international project which plans to develop alternative energy sources to counteract current and future problems with both energy supply and climate change. In your write-up, indicate (a) the causes of the current and future problems, and (b) a schedule for the future plans. Draw a Gantt chart to illustrate your schedule.

7 Reports

1 Statements

Start here 1 Work in pairs. Match the photos A–D with the items in the box.

hand-held metal detector (HHMD) baggage X-ray machine
walk-through metal detector (WTMD) CCTV camera

Reading 2 Read this newspaper article and answer the questions below.

The national airport's security procedures failed an important test last month when a security
5 inspector carried out a surprise security check disguised as a passenger. The *passenger* managed to pass through the security checkpoint carrying a
10 knife, according to sources in the investigation team who spoke to *The Mail* yesterday.

The source explained that before arriving at the airport, the
15 passenger had attached a knife to his lower leg underneath his trousers. At the security checkpoint, he passed once through the walk-through metal
20 detector. When the detector beeped, the security official instructed the passenger to stand aside and told him to raise his arms for a manual search.

25 But the passenger ignored the instruction and told the official that he had recently had surgery on his jaw. He explained that the surgeon had put a metal plate
30 inside his jaw, which made the metal detector beep. Although the official was trained to search everyone after an alarm from the detector, he told the passenger
35 that he could go. The passenger then walked quickly away from the security checkpoint into the concourse. The official's supervisor saw this incident,
40 and immediately followed the passenger into the waiting area, where he ordered him to stop. Then the passenger informed the supervisor that he was a security
45 inspector.

1 Who broke the security procedures: the official and/or his supervisor?
2 What sound did the detector give?
3 Why is the word passenger in italics (line 7)?
4 What does *this incident* (line 39) refer to?

3 Suggest the exact words spoken by the people in each incident.

Examples: *1 'Stand aside(, please).' or 'Would you mind standing aside, please?'*

1 the security official instructed the passenger to stand aside
2 and told him to raise his arms for a manual search
3 the passenger told the official that he had recently had surgery on his jaw
4 He explained that the surgeon had put a metal plate inside his jaw, which made the metal detector beep
5 he told the passenger that he could go
6 he ordered him to stop
7 the passenger informed the supervisor that he was a security inspector

Listening

4 In the investigation into this incident, what questions do you think the investigator asked the security official who allowed the 'passenger' to carry a knife through the checkpoint?

5 21 Listen to the interview between the investigator and the official, and compare the investigator's questions with your questions in 4.

6 Listen again. Note at least four details which are different from the newspaper article in 2. (Notice what the official says, and also what he doesn't say.) Discuss your answers with a partner and make notes.

7 Fill in the gaps in the interview. Check your answers in the audio script on page 122.

1 I told him __________ step back, and then I ordered him __________ walk through again.
2 My supervisor asked me what __________ happened.
3 I told him that the passenger __________ carrying any metal.
4 A: What were your exact words?
B: I said, 'He __________ carrying any metal.'

Language

Statement	Reported statement		
"I am a policeman."	He **told me**	(that)	he was a policeman.
"I'm not carrying any metal."	He **informed her**		he wasn't carrying any metal.
"You can go."	He **told her**		she could go.
"I have searched them."	She **reported**		she had searched them.
"I was an inspector."	He **explained**		he had been an inspector.

Instruction	Reported instruction		
"Open your bag."	He **told him**	to	open his bag.
"Don't pass through."	She **instructed him**	not to	pass through.

8 Change these sentences to reported speech. Use a different reporting verb from the box for each sentence.

tell inform order instruct assure confirm explain promise

Examples: *1 The pilot told everyone to leave the plane immediately.*
2 The passenger assured the policeman that he was innocent.

1 The pilot said to everyone, "Please leave the plane immediately."
2 The passenger said to the policeman, "I'm innocent."
3 The security official said to the man, "Walk through the gantry."
4 The pilot said, "I can confirm that the plane is safe."
5 The policeman said to the passenger, "Come with me."
6 The security manager said to his staff, "There was a security incident this morning."
7 The security official said, "I will be more careful in future."
8 The inspector said, "I pretended to be a passenger."

Speaking

9 Work in pairs. Take turns to role-play the interview between the investigator and the security official. Use all the information from this lesson.

2 Incidents

Start here 1 Work in pairs. Read this case study quickly and discuss the question at the end. Make a note of your solution.

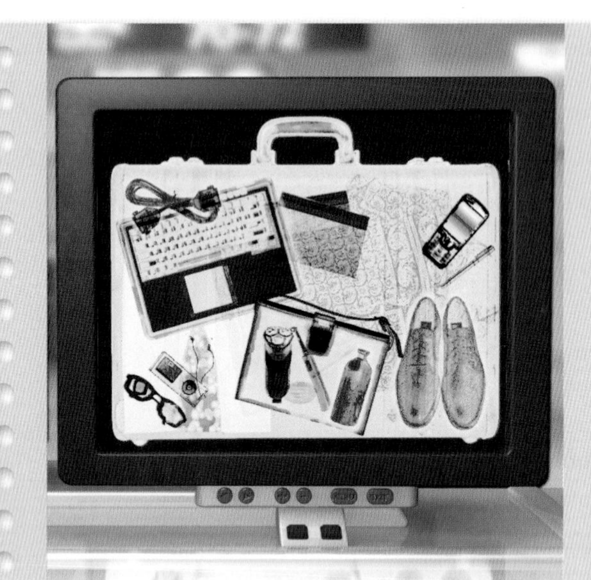

A passenger placed his bag onto the rollers of the X-ray machine. While his bag was going through the machine, he walked through the walk-through metal detector (WTMD) and the alarm sounded. The security official prepared to search him.

While the passenger was passing through the WTMD, and while the alarm was still sounding, a second official was studying a screen on a display unit about eight metres away. He shouted to the first official, and pointed at the passenger's left foot. The first official then told the passenger to take off his left shoe, and found a small knife inside.

QUESTION: *How did the second official know that the knife was in the passenger's left shoe?*

Reading 2 Scan this product review quickly to check your solution to the question in 1.

The Detektit 200 airport metal detector

The Detektit 200 walk-through metal detector (WTMD) detects all metal objects, but ignores harmless ones. It has eight separate detection zones.

The detection zone shows the *vertical* location of the metal object (for example, 0.5 m above the ground), and the strength of the signal shows the *horizontal* location (for example, left leg or right leg). In other words, it can detect the exact location of the items on the passenger's body.

The detector comes with a hand-held remote display screen which can operate at a distance of up to 65 m from the detector. The screen shows the shape of a human body, and displays a flashing light at the exact location of the metal object on the passenger's body.

TECHNICAL DATA

The Detektit 200 is based on the latest pulse-induction technology. On one side of the WTMD, there are eight separate coils of wire, each one at a different height. Short bursts of electric current are sent through each coil. Each pulse generates a magnetic field in the coil.

If the magnetic field meets a metal object (such as a knife or gun), it produces an electrical current in the object, which acts as a resistor. This generates a magnetic field around the metal object.

Meanwhile, as the magnetic field of the coil collapses, it creates a second, very short electric current (called the *reflected pulse*).

The magnetic field from the metal object interferes with the magnetic field from the reflected pulse. This makes the reflected pulse take longer to disappear.

The length of the reflected pulse is compared with the expected length. If the reflected pulse takes too long to die out, the signal is converted to a direct current (DC) voltage.

The DC voltage increases as the metal object moves closer to the coil, and decreases as the object moves away from the coil. The voltage is amplified and controls an audio circuit, which becomes louder and higher in pitch as the metal object gets closer.

3 Read the product review in detail and put these illustrations in the correct sequence.

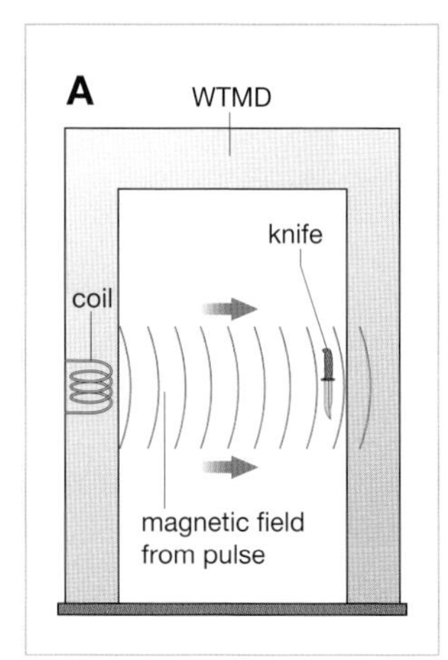

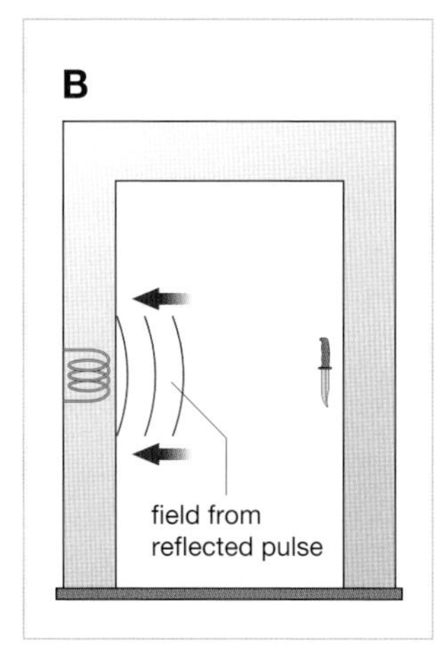

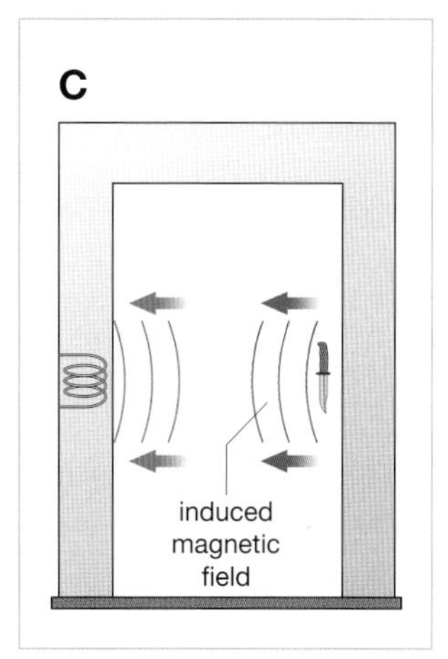

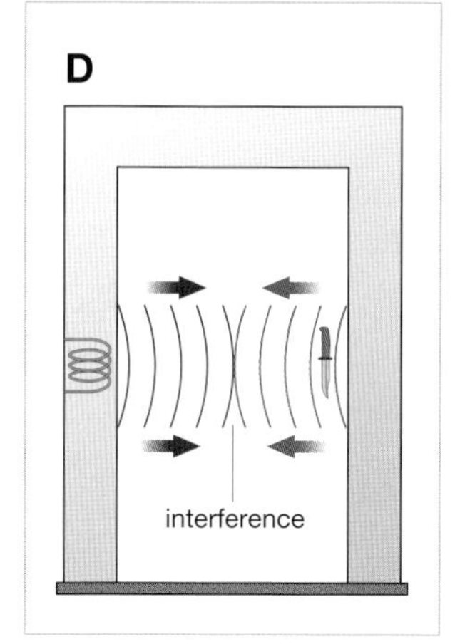

4 Find words in the text that mean the same as the following.

1 metal wires in a continuous circular or spiral shape
2 short burst of electricity
3 area where force (such as magnetism) can be detected
4 a component that stops or slows down the flow of electric current
5 increased

5 Answer these questions about the product review.

1 What process creates the magnetic field around the metal object (such as a knife)?
2 What is the reflected pulse caused by?
3 If there is a metal object, how does this affect the length of the reflected pulse?
4 If the reflected pulse is longer than expected, what effect does this have on the volume of the alarm?

Language

Situation in progress		Short action	
While / When / As	his case **was going** through the X-ray machine,	the alarm **sounded**.	
His case **was going** through the X-ray machine		when	the alarm **sounded**.

Situation in progress		Situation in progress
While / As	the cases **were going** through the X-ray machine,	the officials **were studying** the screen.

6 Make similar sentences from these notes.

Examples: 1 While Team A were searching for the man, Team B were looking for the bag.
2 The official was looking at the passenger when he ran away.

1 08.13–08.32: Team A search for man 08.13–08.32: Team B look for bag
2 14.39–14.41: Official looks at passenger 14.41: Passenger runs away
3 06.20–06.40: Passengers leave plane 06.20–06.40: Staff take bags off plane
4 11.35–12.10: Police patrol Terminal 1 12.10: They suddenly see thief
5 16.05: The passenger becomes ill 15.55–16.05: Flight BA455 takes off
6 20.15–20.25: Flight KL203 lands 20.15–20.25: Flight JL111 takes off.
7 16.30–17.30: Passengers wait in terminal 17.00: Plane leaves with no one on board
8 13.47: The security incident happens / All day: Thousands of people use airport

Speaking 7 Choose an international event that you remember, e.g. Barack Obama became US President on 20th January, 2009. Tell the class about your situation – where you were living, what you were doing – when it happened or when you heard about it.

3 Progress

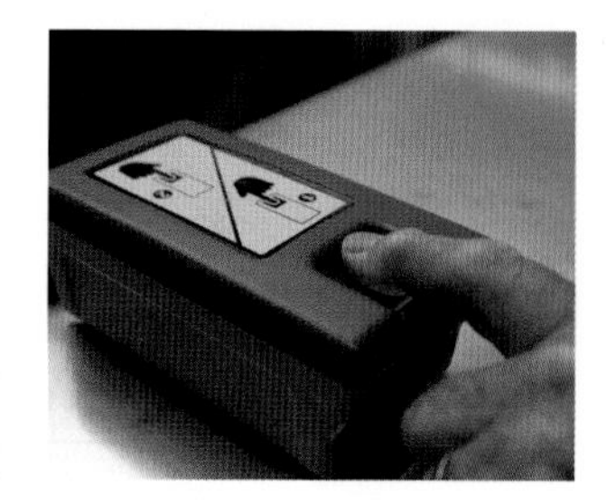

Start here **1** Work in pairs. Which of these methods do you think would be more secure, and why.

There are two main types of fingerprint scanner. *Optical* scanners take a photograph of the finger. *Capacitive* fingerprint scanners use sensors to detect the ridges of the finger.

Scanning **2** Practise your speed reading. Look for the information you need on the SPEED SEARCH pages (116–117). Try to be first to complete these statements.

1 The earliest fingerprints were made in __________, about __________ years ago.
2 Over __________% of people have the 'loop' fingerprint pattern.
3 Identical twins (*share / do not share*) the same fingerprints.
4 Fingerprint evidence convicted the first criminal in the year __________.

Listening **3** 22 Listen to this progress report and complete the checklist. Write D for job done, I for job in progress or P for job planned.

Task: researching …

… passwords	☐	… optical scanning	☐
… pin numbers	☐	… capacitive scanning	☐
… voice recognition	☐	… iris scanning	☐
… fingerprint scanning	☐		

4 Which security systems will Bob not recommend at the end of his research?

5 What is Bob's answer to the question in 1?

Vocabulary **6** Work in pairs. Match these words with 1–7 in diagrams A and B below.

capacitor conductor battery earth resistor switch terminal

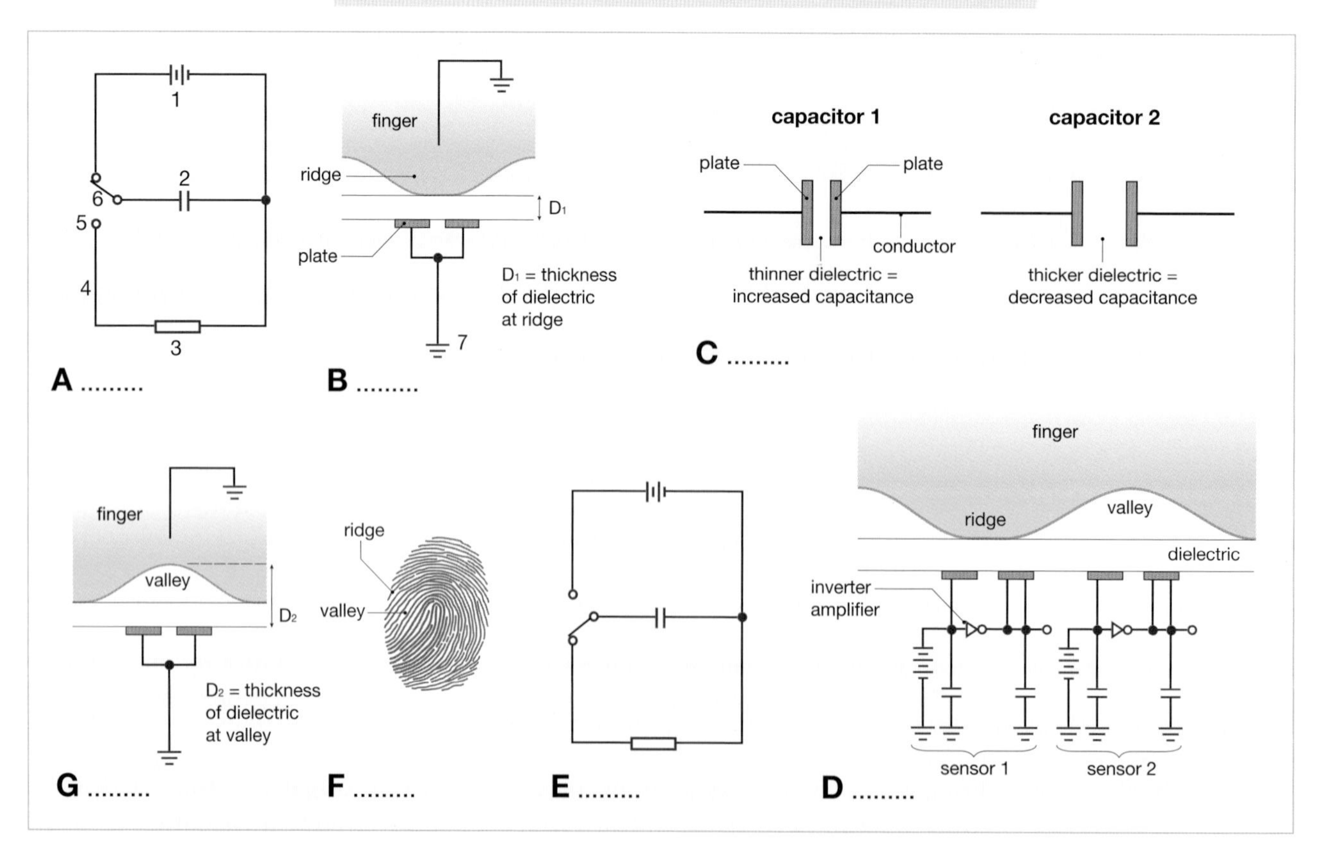

Reading **7** Read this transcript of a lecture, and match the figure numbers 1–7 to the diagrams A–G in 6.

Let's look at the fingerprint scanner in some detail. Basically the finger acts as one plate of a capacitor, and the sensor acts as the other plate. The finger and the sensor (like the two plates of a capacitor) hold an electric charge.

To help understand this let's look at a simple capacitor circuit. The capacitor consists of two metal plates separated by a special non-conductive material called a dielectric. When the switch connects up the charge circuit (see Fig. 1), the battery charges up the capacitor, which will hold its charge for a period of time. Later, when the switch disconnects the battery, but connects up the discharge circuit (see Fig. 2), the capacitor then discharges its load. That's how a camera flashgun works, for example.

The capacitance is the amount of charge which can be held by the capacitor. If the distance between the plates is increased, the capacitance will decrease, as you can see in Fig. 3.

To see how this works with the fingerprint scanner, we can think of the fingerprint as a pattern of ridges and valleys, as in Figure 4.

Now, below each ridge and each valley of the fingerprint there is a single sensor cell, which is part of a microchip.

The single cell actually has a pair of conductor plates, but you should think of the pair of plates as a single plate of the capacitor. So as I said at the beginning, the ridge (or valley) acts as one plate of the capacitor, and the pair of plates in the sensor cell acts as the other plate.

In the case of Figure 5, where the ridge is above the cell, the dielectric, that is, the space between the two plates of the capacitor, is quite thin. So the capacitance under the ridge is quite high.

However, in the case of Figure 6, where the valley is above the cell, the dielectric includes a pocket of air. So the dielectric here is thicker. That means that the capacitance under the valley is quite low.

So all the cell has to do is to measure these differences in capacitance. And how does it do that?

As you can see in Figure 7, the capacitor circuit is connected to another circuit which passes through a special kind of amplifier.

When current passes through the amplifier, it is modified by the capacitance of the capacitor circuit. A high capacitance will change the current to one voltage, while a lower capacitance will change the current to a different voltage.

So under a ridge, there will be one voltage output from the amplifier, and under a valley there will be a different voltage.

The processor then reads the voltage output of all the cells in the fingerprint scanner and builds up a total picture of the ridges and valleys in the fingerprint.

fingerprint scanner capacitor sensor non-conductive dielectric capacitance ridge valley microchip amplifier 23

8 Are these statements *true* (T) or *false* (F)? Correct the false ones.

1 In a capacitor, a dielectric separates two conductor plates.
2 When you connect the power source to a capacitor, the capacitor is discharged.
3 If you widen the dielectric, you increase the capacitance.
4 The capacitance under a valley is higher than the capacitance under a ridge.

Writing **9** Work in small groups. Prepare a talk explaining how capacitive fingerprint scanners work.

Write the main points about each visual (in the correct order: Figs 1–7 in 7) on cards.

Speaking **10** With your group, give the talk to the whole class, and answer questions.

8 Projects

1 Spar

Start here

1 Work in pairs. Match the parts of the illustration with the labels in the box.

spar mooring line topside pumping station riser pipeline tree

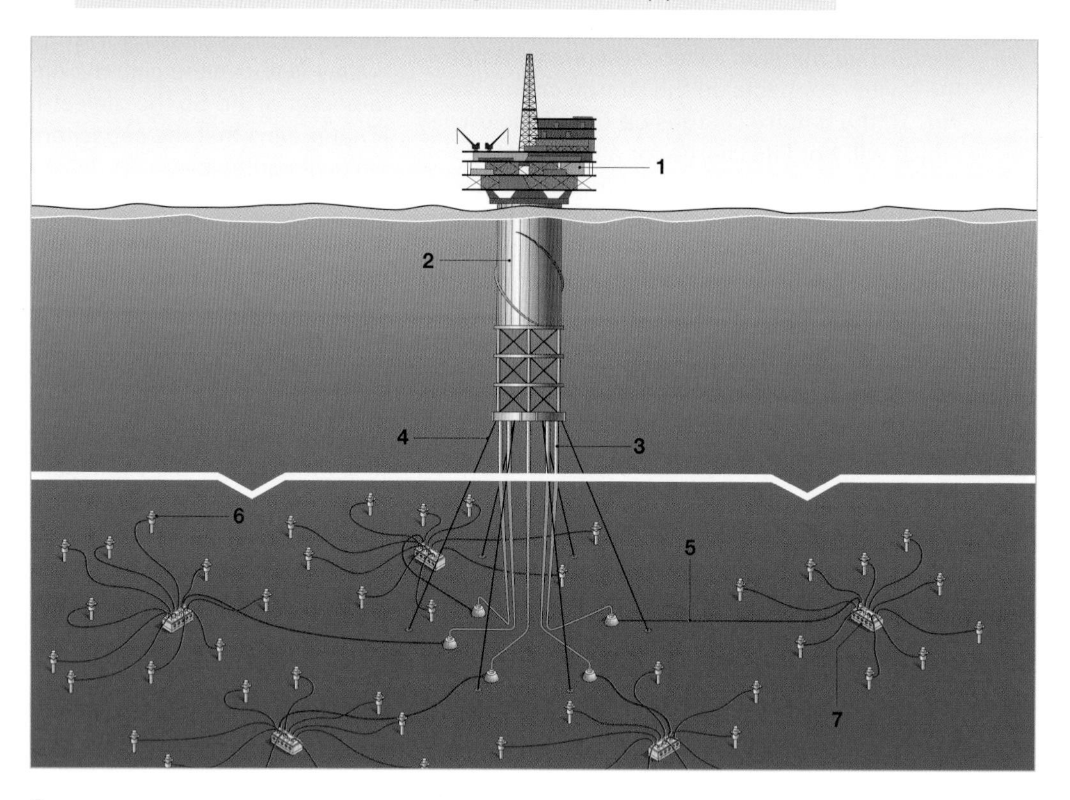

2 Discuss in pairs. Where is the deepest spar in the world? How deep (approximately) do you think it is?

Listening

3 ▶ 24 Listen to these news items and check your answers to 1 and 2.

4 Listen again, and complete the specifications chart.

Perdido Spar offshore oil platform: Specifications	
Total length of spar	________ metres
Diameter of spar	________ metres
Height of spar platform above seabed	________ metres
Weight of spar	approx ________ tonnes
Number of mooring lines	________
Number of risers	________
Total number of wells	________
Average oil and gas production in first year	________ b/d

Note: *b/d* = barrels per day

5 Complete these sentences from the news items with the correct form of the verbs in the box.

produce lay complete build drill tow secure fit

1 Twenty-two oil wells ________________ below the Perdido Spar oil platform
2 The first oil well under the Perdido Spar ________________ .
3 The Perdido Spar ________________ to the seabed.
4 More than 46 million barrels of oil ________________ by the Perdido Spar.
5 The topside ________________ to the top of the Perdido Spar.
6 Five risers and a pumping station ________________ below the world's deepest spar.
7 A huge pipeline network ________________ under the Perdido Spar.
8 The Perdido Spar ________________ to its site in the Gulf of Mexico.

6 Write numbers 1–8 (the numbers of the sentences in 5) in the correct boxes on the timeline. Check your answers in the audio script on page 123.

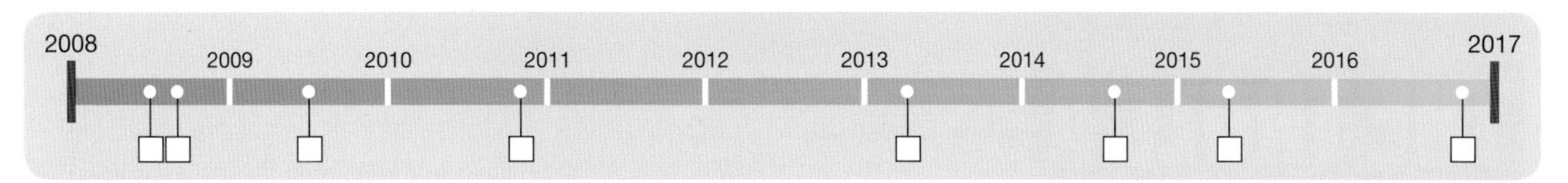

Language

Present perfect passive	Past simple passive
The topside ***has been fitted*** *to the spar.*	*The topside* ***was fitted*** *to the spar in 2009.*
Five risers ***have been built****.*	*The risers* ***were built*** *three years ago.*

7 Complete this dialogue. Use the correct passive form of the verbs in brackets.

A: (1) *Has the spar been towed* (the spar / tow) to its site yet?
B: Yes, it has. (2) ____________ (It / also / secure) to the seabed.
A: Really? When (3) *was it brought* (it / bring) to the site?
B: In June last year. And it (4) ____________ (fix) to the seabed that August.
A: Good. What about the topside? (5) ____________ (it / fit) to the spar yet?
B: Yes, it has. And a 24-man crew (6) ____________ (already / install) on the rig.
A: That's great. When (7) ____________ (the topside / attach) to the spar?
B: In June this year. And the crew (8) ____________ (take) to the rig in August.
A: Great. (9) ____________ (any pipelines / lay) along the seabed yet?
B: No, but the first risers (10) ____________ (already / lower) to the seabed.
A: Really? When (11) ____________ (that / do)?
B: The risers (12) ____________ (drop) to the seabed about two weeks ago.

Speaking 8 Practise the dialogue in 7 with a partner.

Scanning 9 Practise your speed reading. Look for the information you need on the SPEED SEARCH pages (116–117). Try to be first to complete these statements.

1 The Perdido Spar is ______ km south of Houston, Texas.
2 The total operating weight of the spar is ______ tonnes.
3 Oil was first discovered under the sea at the Perdido site in the year ______.
4 The depth of the water around the spar ranges from 2,280 to ______ m.

2 Platform

Start here

1 Work in pairs to answer these questions.

1. What is the structure in the photo on the right used for? (Here it is shown next to the Eiffel Tower in France to illustrate its height and size.)
2. What world record do you think the structure holds?

Reading

2 Read the text and check your answers to 1. Complete the text with the statistics in the box. Then discuss your answers with a partner or in a small group.

656,000 t 50 mm 1995 303 m 100,000 t 20 min 1.3 trillion m³
472 m 2,000 200 km 35 m 245,000 m³ 25

THE TROLL A

offshore oil and gas drilling platform, at a total height of (1) *472 m*, is the tallest construction which has ever been transported from one location to another. The structure is also one of the largest and most complex engineering projects in history.

It was built over a period of four years, using a workforce of about (2) ____________, and it was first deployed in (3) ____________ to produce natural gas from the enormous Troll oil field off the coast of Norway.

The total weight of the reinforced concrete structure is (4) ____________, which includes (5) ____________ of reinforcing steel. The volume of the concrete alone (without the steel) is (6) ____________.

The platform is located in sea water which is (7) ____________ deep at that point, and its legs are sunk (8) ____________ into the seabed in order to provide sufficient stability.

The structure has four legs made of reinforced concrete. Each leg has an elevator that takes over nine minutes to travel from the platform above the waves to the seabed.

The legs are hollow and have walls one metre thick. Their shape is a combination of cylindrical and conical, but the diameter of each cylinder is wider at the top and bottom.

The legs were built by pouring concrete continuously, by means of a process called *slip-forming*, in order to give them maximum strength and make them able to withstand intense pressure.

This process was a lengthy one, taking (9) ____________ to add only (10) ____________ to the height of each leg. The *form* (which forms the wet concrete into the correct shape) moved (or slipped) slowly up the tower, pushed by hydraulic cylinders, while the concrete was being poured. The continuous pour continued day and night without a break for several months until each leg was completely built.

When the legs were finished, the platform was fixed to the top, and then the whole structure was towed over the sea a distance of (11) ____________ from the construction site to its final location on the Troll gas field, 80 km north-west of Bergen in Norway.

The Troll platform is connected to electricity supplies on the mainland by means of a sub-sea electrical cable.

The Troll oil and gas field is one of the largest offshore fields in the world with recoverable reserves of approximately (12) ____________ of gas. Today, the field produces up to 100 million m³ of natural gas every day.

Slip-forming in progress. The form is at the top, covered in white fabric.

reinforced concrete structure   slip-forming   hydraulic cylinders
continuous pour   sub-sea electrical cable   recoverable reserves   26

3 What do these words refer to in the text in 2?

1 The structure (line 5)
2 Their (line 21)
3 them (line 25)
4 This process (line 27)

4 Find words or phrases in the text with the same meaning.

1 group of workers
2 used or put into action
3 strengthened
4 very powerful
5 long (in time)
6 ongoing, not stopping
7 flow
8 below the seabed

Language

The spar was secured	**method**
	by fastening it to the seabed with nine cables.
	using / by means of nine cables.
	purpose
	(in order) to prevent it from moving around in heavy seas.

5 Match these phrases from the text in 2 with their language function. Write M (method) or P (purpose) next to each phrase.

1 to produce natural gas from the enormous Troll oil field (line 9) ☐
2 in order to provide sufficient stability (line 17) ☐
3 by pouring concrete continuously (line 24) ☐
4 by means of a process called *slip-forming* (line 25) ☐
5 in order to give them maximum strength (line 25) ☐
6 by means of a sub-sea electrical cable (line 39) ☐

6 Write single sentences in the past to express the same meanings as the notes below. Use language from the table above.

Example: *1 The highest structure in the world was built using 330,000 m³ of concrete.*

1 Highest structure in world: built. Method: use 330,000 m³ of concrete.
2 Concrete: made in a single pour. Purpose: prevent cracks from forming.
3 Topside: fitted to top of Perdido Spar. Method: deploy hydraulic winches
4 Pumping station: installed below spar. Purpose: separate the oil from the gas
5 Spar: secured to the seabed. Method: use nine mooring lines
6 Five risers: built below spar. Purpose: carry oil and gas up to platform.

Vocabulary

7 Complete sentences choosing the correct form in brackets.

Note:
- *reinforced* concrete = concrete *which has been reinforced* (passive meaning)
- *reinforcing* steel = steel which *reinforces* something (active meaning)

1 A rebar is a (reinforced / reinforcing) steel bar that strengthens concrete.
2 This (reinforced / reinforcing) plastic has been strengthened by carbon.
3 They are using (pressurised / pressurising) oxygen in those tanks. The gas was compressed before going into the tanks.
4 We need to fit a (pressurised / pressurising) valve onto this pipe, to create some pressure in the system.

3 Drilling

Start here

1 Work in pairs. Match the parts of the oil rig with the words in the box.

hook swivel turntable drill pipe
drill bit winch cable travelling block

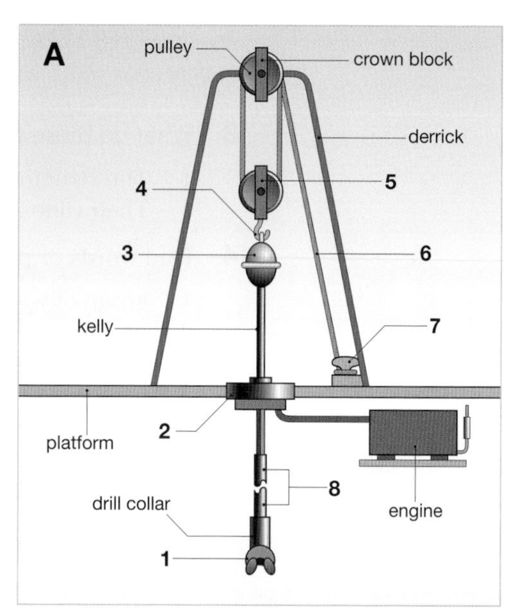

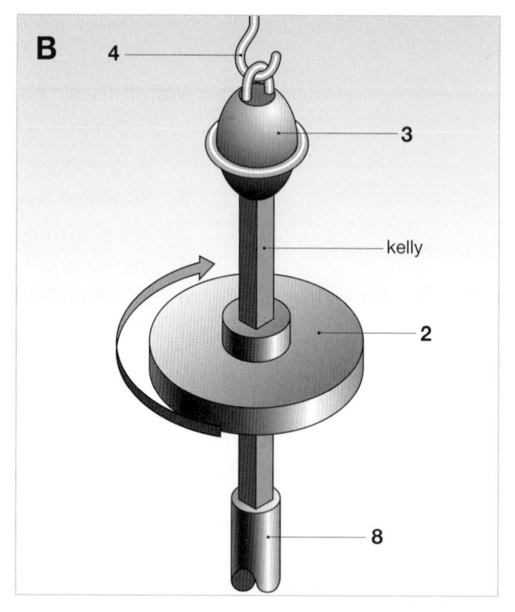

2 Discuss with your partner which parts of the rig (a) ***rotate***, (b) ***move up and down*** and (c) do both. Write **R** (***rotate***), **U** (***up and down***) or **RU** (***both***) next to the words in the box in 1.

3 ▶ 27 Listen to this description of the diagrams in 1, and check your answers to 1 and 2.

Listening

4 ▶ 28 Listen to this TV interview with the oil worker and put these stages into the correct order.

Preparing to drill an oil well

A Check the mud hose for leaks. ☐
B Lower the drill string through the rotary table. ☐
C Switch on the mud pump. ☐
D Slide the drill collar over the drill pipe. ☐
E Switch on the power to the rotary table. ☐
F Lower the drill bit to the rock layer. ☐
G Attach the drill pipe to the drill bit. 1
H Start drilling. ☐
I Fix the kelly to the drill pipe. ☐
J Lower the drill string into the well hole. ☐

5 The driller in the interview in 4 used both active and passive verbs when describing his actions. Mark each stage in 4 either A (*active*) or P (*passive*) according to what you remember from the audio.

6 Listen again to check your answers. Say why the driller used active or passive verbs.

7 Complete the interviewer's words.

1 Sorry, could you just explain ____________________ *made up*?

2 Excuse me, ____________________ *string*?

3 Sorry, ____________________. It's the noise. Could ____________________?

4 Sorry to ____________ you, but ____________________ *tripped in* mean?

8 Work in pairs. Describe how the well was drilled. Use the past simple passive. Take turns, with A describing the first half and B describing the second half of the process. Practise the language in 7 by asking each other to explain or repeat information.

Begin: *First, the drill pipe was attached to the drill bit. Next, …*

Speaking

9 Work in a team of four. Study and read your own material individually (as explained below). Then meet as a group to discuss how the complete process works. Make notes about the complete process.

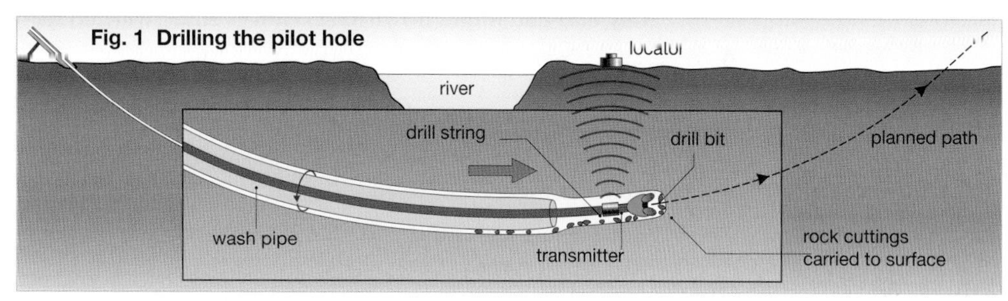

Fig. 1 Drilling the pilot hole

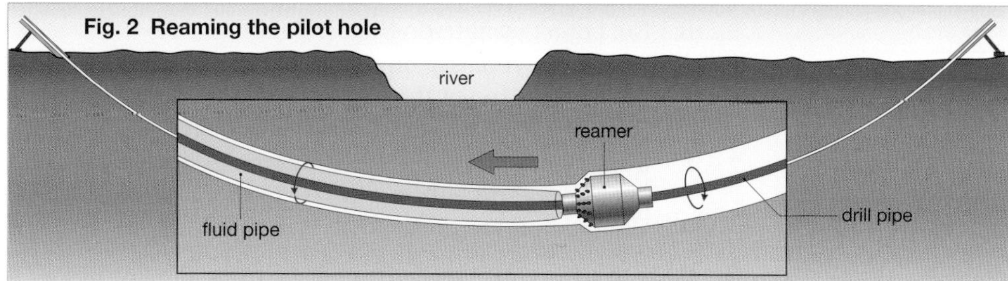

Fig. 2 Reaming the pilot hole

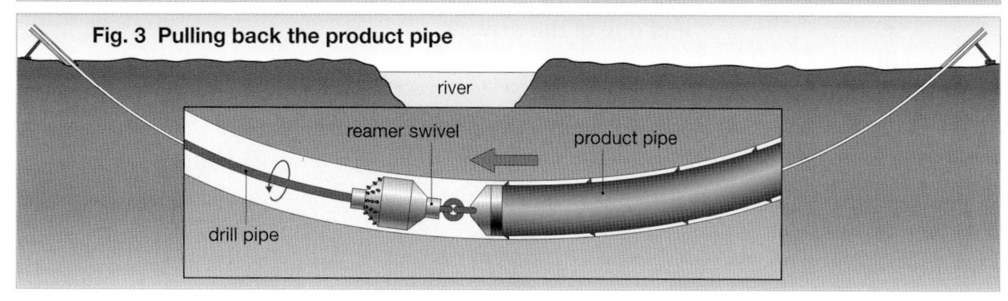

Fig. 3 Pulling back the product pipe

The purpose of this process is to drill a hole below an obstruction such as a lake, road or river so that a large pipe called the product pipe (for carrying water, sewage, electricity, etc.) can pass below the obstruction. The process consists of three stages.

Student A: Study Fig. 1 on this page, and read Fact sheet 1A on page 109.
Student B: Also study Fig. 1 on this page, and read Fact sheet 1B on page 111.
Student C: Study Fig. 2 on this page, and read Fact sheet 2 on page 112.
Student D: Study Fig. 3 on this page, and read Fact sheet 3 on page 114.

Writing

10 Work individually. Complete this report of how a tunnel was drilled and a pipe installed below the River Avon. Use the past simple passive where appropriate. Refer to the notes from your meeting in 9.

Drilling the pilot hole

First a drill bit with a narrow diameter was attached to a drill string. Then a narrow pilot hole was drilled …

Review Unit D

1 Give the exact words spoken.

Example: *1 'Don't fire, but stand ready.'*

1 The commanding officer ordered his troops not to fire but to stand ready.
2 The new recruit explained that he was not a police officer but that he was trained in security procedures.
3 Our new manager announced that he had introduced a new pay system for all the staff, but told us not to worry. He said that our pay would rise by 5%.
4 The IT technician reported to us that he had checked our computers the previous day and had found no viruses. However, he warned us to be careful in future.

2 Write this dialogue as a report. Use the verbs in brackets.

A is an airport security official and B is a male passenger.

Begin: *The official told the passenger to stop and take off his shoes. The passenger agreed to do this.*

A: Excuse me, sir. Could you please stop and take off your shoes. (tell)
B: All right. (agree)
A: Now please put your bag on the rollers. (ask)
B: No, I don't want to do that. (refuse)
A: Please put your bag on the rollers. (ask again)
B: I don't want to put the bag through the X-ray. (say) It contains undeveloped film. (explain) I don't want to damage it. (tell)
A: Don't worry. (advise) The X-ray won't damage your film. (inform) Now please put your bag on the rollers. (instruct)
B: OK. (comply)

3 Work in pairs. Take turns to act the part of a police officer investigating this accident, and one of the drivers.

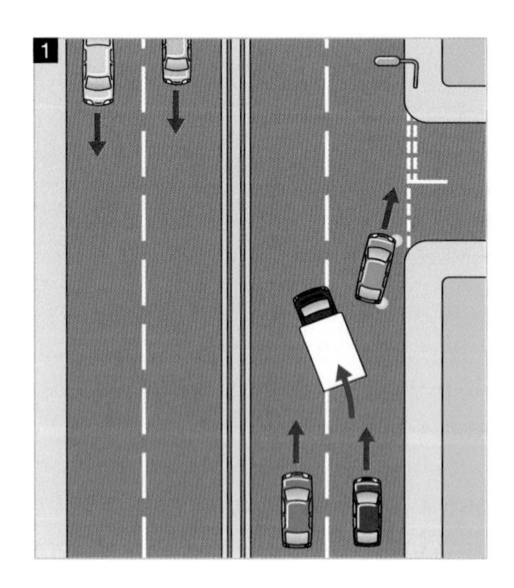

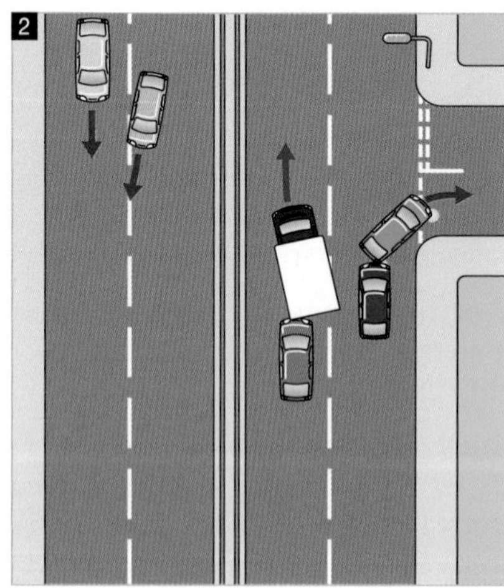

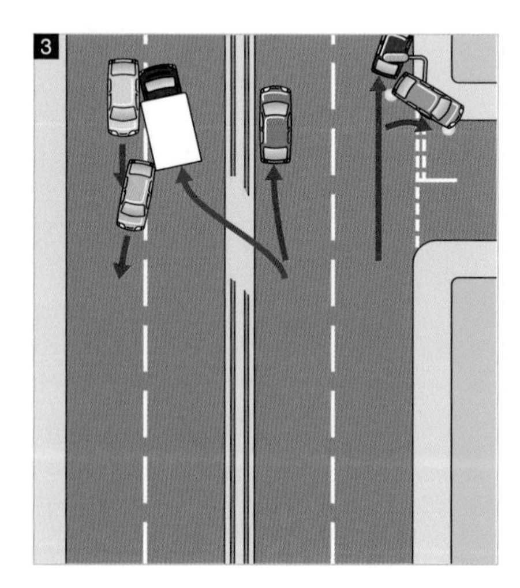

Example: *(A is the police officer, and B is the driver of the white van.)*
A: *What happened? (How did the accident happen? What were you doing?)*
B: *I was following a green car, when suddenly the driver indicated to turn right. As I was moving into the left-hand lane, the red car crashed into ...*

4 You are the police officer. Write a brief report describing what happened in this accident. Use *while / when / as* + past continuous where appropriate.

5 Complete each sentence using two different nouns both related to the verb in brackets.

1. The antenna is a *transmitter*, which is used for the *transmission* of signals to the satellite. (transmit)
2. This hand-held metal ____________ is used for the ____________ of dangerous items such as guns or knives. (detect)
3. You should connect an ____________ to your MP3 player. This will give the music the ____________ which you need. (amplify)
4. This country is a major ____________ of crude oil for the world's refineries. Oil ____________ makes up about 40% of its industry. (produce)
5. If you want us to increase our level of electricity ____________, you'll have to allow us to build a larger ____________. (generate)
6. One way to increase crop yields would be to install a huge ____________ in space. The ____________ of the sun's rays to the dark side of the Earth would almost double the growth of agricultural crops. (reflect)

6 Work in pairs. Study the diagrams and the notes below, and discuss how the device works.

Purpose of machine

- convert electrical current into rotational movement

Scientific principle

- electric current in a wire coil generates a magnetic field in a nearby conductor

Main parts

- permanent magnet
- electromagnet

Parts of electromagnet

- *coil* – wire made of copper – goes around armature – connected to power source via brushes and commutator – carries electrical current – magnetises armature
- *armature* – solid bar made of iron or steel – magnetised by current in coil – becomes electromagnet – rotates around axle – between arms of permanent magnet – north pole of armature repelled (pushed away) by north pole of permanent magnet – north changed to south, south changed to north (by brushes and commutator)
- *brushes* – made of steel – connected to power source – one brush positive charge, other one negative – passes current to commutator and coil
- *commutator* – made of steel – touches brushes as it rotates – allows current to flow into coil – touches different brush (negative then positive) – changes magnetic pole of armature

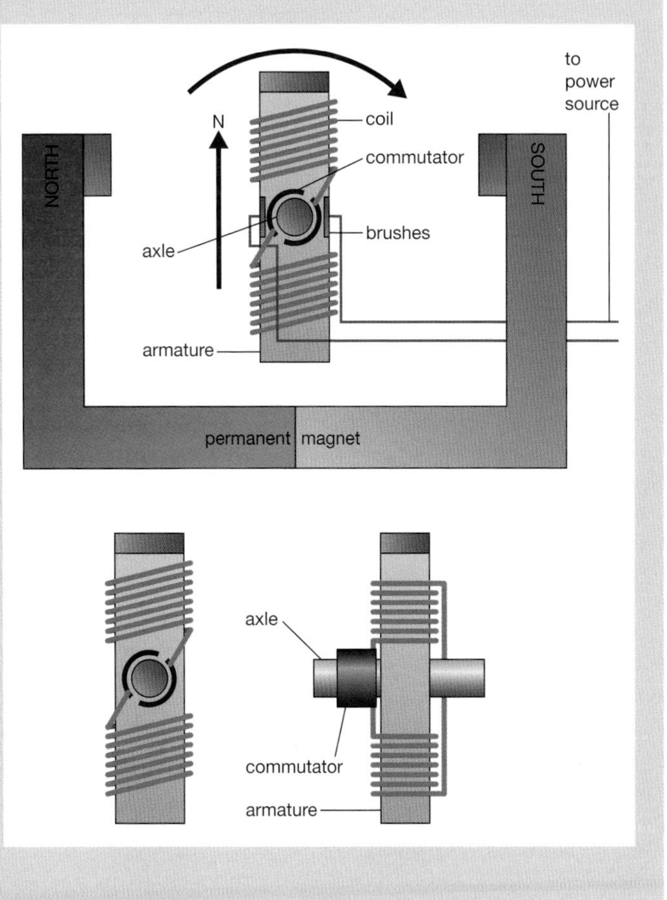

7 Write an explanation of how an electric motor works.

8 Discuss with a partner how the principle used in an electric motor is similar to that used in a metal detector.

9 Complete the sentences using either ***to*** or ***by*** and the correct form of the verb in brackets.

1 The warehouse was secured __________ (install) CCTV cameras.

2 The tunnels have been dug __________ (link) the two sides of the river.

3 The legs of the spar were constructed __________ (pour) concrete into them continuously.

4 The hand-held metal detector was used __________ (search) the passenger's body.

5 A swivel is used __________ (allow) the string to rotate freely.

6 The walk-through metal detector works __________ (generate) a magnetic field in the metal object.

10 Write sentences from these notes explaining what has been done. Use the present perfect passive.

Example: 1 *The spar has been transported to its site in the Atlantic Ocean.*

1 *transport spar to site in Atlantic Ocean*

2 *moor spar to sea floor using cables*

3 *attach topside to top of spar*

4 *set up pipeline network below spar*

5 *dig foundations for pumping stations below seabed*

6 *construct risers and install pumping stations*

11 Complete this dialogue. Use the passive form of the present perfect or past simple.

A: Have the walls (1) <u>*been plastered*</u> yet? (plaster)

B: No not yet. But the wiring (2) __________, so the plastering can start soon. (already / install)

A: Oh that's good. When (3) __________? (the wiring / put in)

B: It (4) __________ yesterday. (finish)

A: Right, and (5) __________ (paints / buy) yet?

B: Not yet, but they (6) __________ (choose) by the customer, and they (7) __________ (already / order).

A: When (8) __________ (the order / make)?

B: The order (9) __________ (send) to the suppliers this morning.

12 Make some notes about a simple process you know about. Then explain it to a partner. Take turns. When your partner is talking, interrupt them politely (pretending you can't hear, or didn't understand) using the correct forms of these words and phrases.

sorry excuse could you repeat would you mind explain the meaning of what do you mean by interrupt repeat say again What does ... mean? I didn't catch that

13 You are a technical journalist. Imagine that the bridge below has recently been completed. Write questions to get the information on the factsheet.

EXTREME ENGINEERING FACTSHEET

NAME: Sheikh Rashid bin Saeed Crossing (also called Sixth Crossing)
LOCATION: Dubai, United Arab Emirates
PURPOSE: to connect localities of Al Jaddaf and Bur Dubai
ESTIMATED COST: $817 million
NUMBER OF ARCHES: two
HEIGHT OF LARGER ARCH: 205 m
RECORD MADE: the world's tallest arch bridge on a bed of sand
RECORD BROKEN: twice the height of record-holding Lupu Bridge, Shanghai, China

ACHIEVEMENT: extreme engineering – built on a bed of sand, instead of rock
LENGTH OF BRIDGE: 1.6 km
HEIGHT OF DECK: 15 m above water
AMOUNT OF STEEL USED: 140,000 t
FOUNDATIONS: 200 steel-reinforced-concrete piles
DEPTH OF FOUNDATION PILES: 40 m
METHOD OF PILE CONSTRUCTION: holes 2 m wide drilled 40 m into sand; holes filled with steel-reinforced concrete

14 Work in pairs. Practise asking and answering the questions you wrote in 13.

15 Work individually. Write a news item for the bridge (as if it has just been completed) for a technical magazine. Use all the information in the factsheet. If possible, before you start, find out more information about the bridge using an internet search engine.

Projects **16** Research one of the following:

1 A security or health and safety incident which affected transport by land, sea or air, or which had a serious effect on your industry. Write a description of the background to the incident, what actually happened, the reasons for the incident and the action that was taken afterwards.
2 A large project which is nearing completion in your technical or industrial field. Write the news report which will be read out when the project has been completed. Explain the methods and purposes involved in the project.

9 Design

1 Inventions

Start here

1 Work in pairs. Discuss the design of this vehicle and compare it with (a) a normal motorbike and (b) a sports car. Consider the features in the box and any others which you think are important. Make notes of your ideas.

number of wheels   speed   acceleration
braking power   ease of turning
stability (on rough roads)
stability (when it turns)   storage space
comfort   excitement   safety

The Can-Am Spyder roadster

Listening

2 29 Listen to this test report on the performance of the Spyder against the Zoomster XL motorcycle. Tick the correct statements in the checklist below.

The Spyder ...

1. has brakes which are a great deal more powerful than the Zoomster's ☐
2. reaches the same maximum speed as the Zoomster ☐
3. accelerates as quickly as the Zoomster ☐
4. is much more difficult to steer than the Zoomster ☐
5. has much better suspension than the Zoomster on a bumpy road ☐
6. is a lot less stable than the Zoomster when taking a bend at speed ☐
7. is slightly less safe than the Zoomster ☐
8. is about three times more expensive than the Zoomster ☐

3 Listen again and correct the statements in 2.

Language

This motorbike is	far much a great deal a lot slightly a little	more expensive	than	that one.
This car accelerates		less quickly more slowly		

We can modify a comparison in a **more specific** way, like this:

The new hard drive is	three times ten percent a third 3 MB	larger		than	the old one.
This engine is	twice three times half	as	powerful	as	that one.

4 Compare the differences between these two types of light bulb.

	CFL bulb	Incandescent bulb
Energy input (watts)	13	60
Light output (lumens)	810	830
Useful life (hours)	10,000	1,500
# bulbs for 10,000 hours	1	6.7
Bulb costs	1 @ €2.7 = €2.7	6.7 @ €0.22 = €1.5
Electricity used (kW hours)	130	600
Electricity cost (@€0.55 per kWh)	€71.5	€330
Total cost (electricity + bulb)	€74.2	€331.5

Cost saved in your lighting bill by using a single CFL bulb over seven years = €257.30.

Examples: *An incandescent bulb consumes much more energy (or many more watts) than a CFL bulb. A CFL bulb lasts more than six times as long as (or six times longer than) an incandescent bulb.*

compact fluorescent light bulb   CFL   incandescent light bulb
13 watts   810 lumens   130 kW hours   @€0.55 per kWh   30

Speaking 5 Work in pairs. You are a team of designers. You have designed a new type of light bulb for a design competition you want to enter. Discuss and plan your proposal. Use information at the back of the book, and from the table in 4 above.

Student A: Read your information on page 109.
Student B: Read your information on page 110.

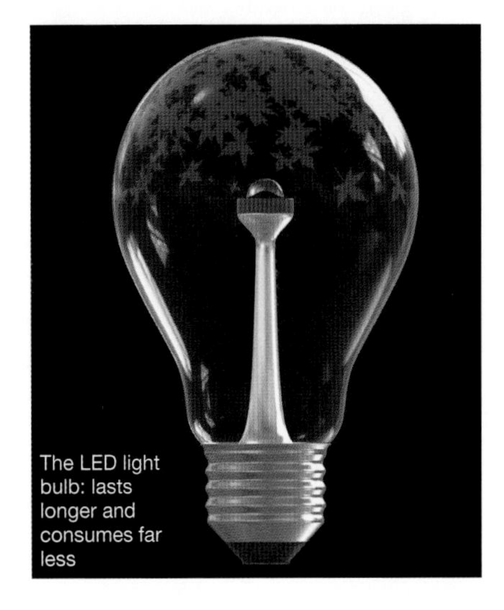

The LED light bulb: lasts longer and consumes far less

Writing 6 In pairs, write your proposal for the design competition. Include the following:

- a short explanation of how your design meets a need in society today
- a comparison of your design with similar products, and an explanation of how it improves on them

Task 7 In pairs, discuss how to improve a product which you use regularly in your work or in your daily life.

For example, you could think about your mobile phone or some computer software or hardware that you use every day, or an appliance in the building where you work or study. How could you improve its design, so that it works better for you?

8 Write the proposal. Explain how your product will compare with other products on the market.

9 Present your proposal to the class, and answer any questions.

2 Buildings

Start here

1 Work in pairs and discuss these questions.

1 What are the names of these buildings?
2 Where are they located?
3 Why do architects say that they have the same structural design?

The answers are on page 111.

2 Imagine the base (or plan) of each building. Which one is roughly (1) ***circular*** (2) ***oval*** and (3) ***rectangular*** in shape?

3 In pairs, take turns to describe each building briefly (without naming or identifying it). Try to identify each one from your partner's description. Use some of these words and phrases when describing the shapes of the buildings, and the shapes of their bases.

straight curved zigzag cylindrical conical diagonal inclined
at an angle vertical horizontal rectangular oval elliptical tapered

Reading

4 Read these fact sheets and write the correct names of the buildings in 1.

Building # 1: ______________________

HEIGHT: 181.97 m
STOREYS: 46
FLOOR AREA: 80,000 m^2
SHAPE: approximately rectangular prism; its vertical edges have a zigzag shape; floor virtually rectangular; uses 21% less steel than a standard design
ENERGY CONSUMPTION: about 25% less than a standard building

Building # 2: ______________________

HEIGHT: 179.80 m
STOREYS: 41
FLOOR AREA: 47,950 m^2
SHAPE: roughly conical, with curved tapering sides; tower bulges out slightly from its base, reaching its maximum width at the 16th floor; floor plan approximately circular
ENERGY CONSUMPTION: half as much as a normal tower of the same size

Building # 3: ______________________

HEIGHT: 160 m
STOREYS: 35
FLOOR AREA: 50,000 m^2
SHAPE: curved tower leaning westwards 18° off vertical (four times as far as the Leaning Tower of Pisa); believed to be 'the most inclined tower in the world'; floor plan roughly elliptical (oval)
the steel weighs 21,500 t (80% lighter than Beijing's Bird's Nest)

5 Correct the false information in these statements.

1 The Hearst Tower has by far the most extensive (or greatest) floor area of the three buildings, and is also the tallest and has the largest number of storeys. The floor plan is the closest to a rectangular shape. The top of the tower is the most tapered (or pointed) of the three.
2 Capital Gate has the fewest storeys of the three, and is easily the shortest of the three towers, with the least extensive floor area. It has the most curvilinear (or curved) shape of the three, and is the most inclined (or leaning) from the vertical.
3 The Swiss Re Building has the most conical overall shape, and the most circular floor plan of the three. The narrowest point is at the top of the tower, and the widest part is at the base. It is the least tall of the three buildings and has the fewest storeys and the least extensive floor area.

Language

<table>
<tr><td rowspan="3">Hearst Tower has</td><td rowspan="4">easily
by far</td><td rowspan="4">the</td><td>greatest</td><td rowspan="2">floor area</td><td rowspan="3">of</td><td rowspan="3">the three buildings.</td></tr>
<tr><td>most extensive</td></tr>
<tr><td>least circular</td><td>floor plan</td></tr>
<tr><td>Capital Gate is</td><td>most inclined</td><td>tower</td><td>in</td><td>the world.</td></tr>
</table>

6 Describe each building briefly (without naming it) using superlative adjectives. See if the others in the class can identify the buildings.

Example: *This building has by far the largest floor area of the three.*

Speaking 7 Work in small groups. Choose the winner of the award for the best tall building. The short-listed buildings are Hearst Tower, the Gherkin, Capital Gate and the two buildings below. Before you begin, decide on your criteria for the award, for example *beauty, innovation, functionality, low material consumption, low energy consumption.* Make notes of your group's reasons for the decision.

Building # 5:

Bahrain World Trade Centre, Manama, Bahrain: 240 m high / 50 storeys / floor area: 120,000 sq m / floor plan: elliptical in shape / world's first building with built-in wind turbines / systems for reducing and recovering energy / inspired by local 'wind towers' to use wind for cooling / uses sail-like shape to increase wind flow between towers

8 With your group, report back to the class, explaining the reasons for your group's choice of best building.

3 Sites

Start here

1 Work in pairs. Study the two photographs and the site plan below, and identify the two buildings on the site plan.

Building 1: Is it A, B or G? Building 2: Is it C, F or H?

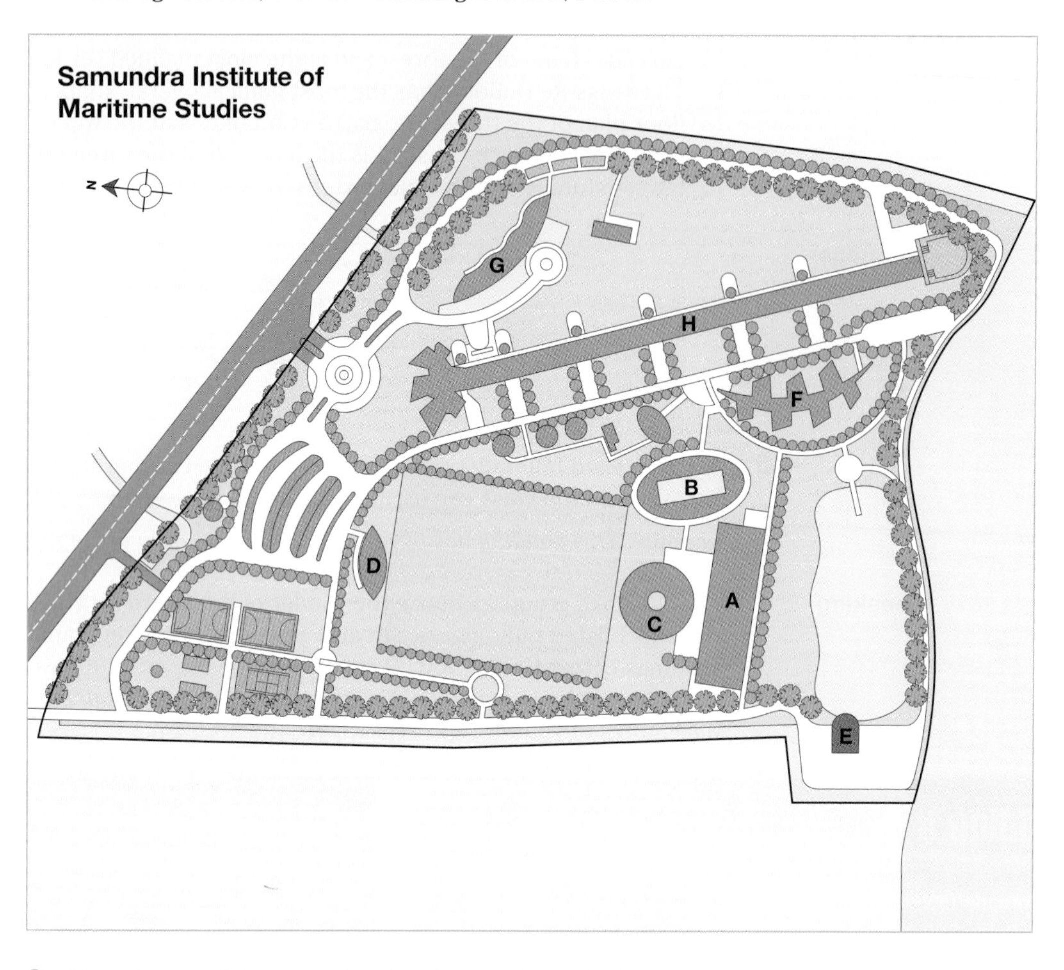

2 Find these buildings on the site plan, and match them with letters A–H.

1 a rectangular building, close to a roughly rectangular lake [A]
2 an oval building, pointed at both ends ☐
3 a slightly curved building which looks like a set of teeth ☐
4 a rectangular swimming pool enclosed in an oval or elliptical building ☐
5 a structure semi-circular at one end and straight at the other end ☐
6 a long, curved building adjacent to (next to) a small curved lake which is tapered at one end ☐
7 a very long narrow rectangular building on the opposite side of the small lake from the long curved building ☐
8 a doughnut-shaped (or ring-shaped) building ☐

doughnut (BrE) = donut (AmE) = ring-shaped

Listening

3 32 Listen and write the letters A–H from the site plan next to the names of these buildings.

1 Administration Building	☐	5 Student Hostels	☐
2 Research Centre	☐	6 Campus Ship	☐
3 Services Building	☐	7 Workshop	☐
4 Academic Block	☐	8 Swimming Pool	☐

Vocabulary 4 Match these sketches with the photos in 1.

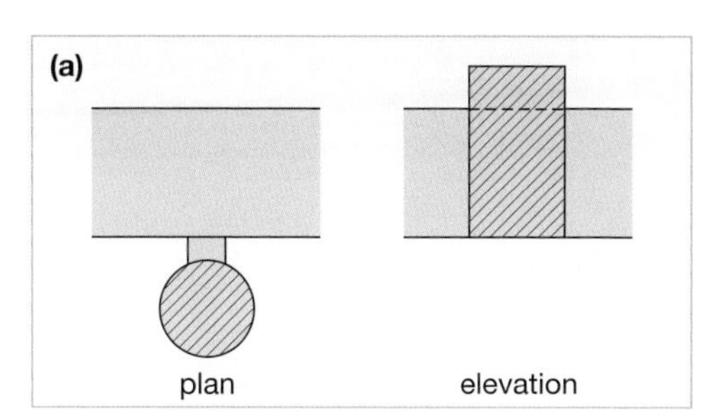

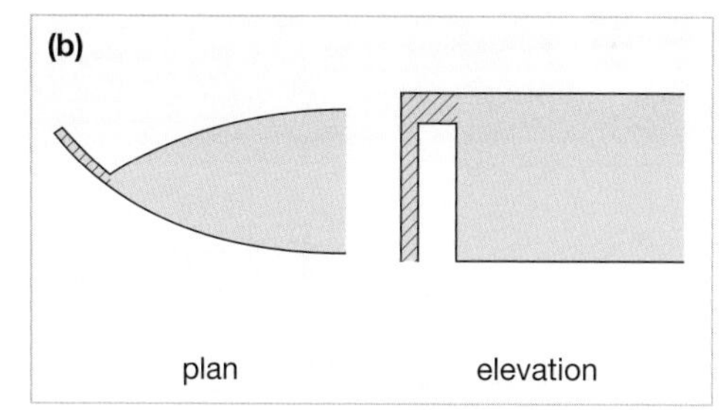

5 Complete the text using the words/phrases in the box. Use the illustrations to help you.

hatched tapered elevation perpendicular horizontal solid plan broken arch

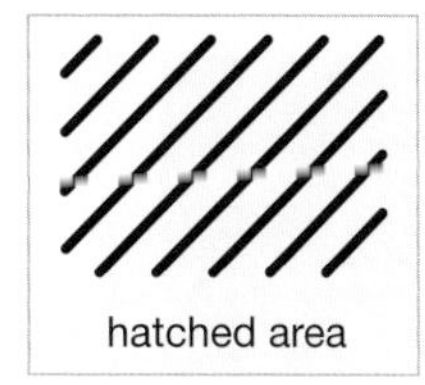

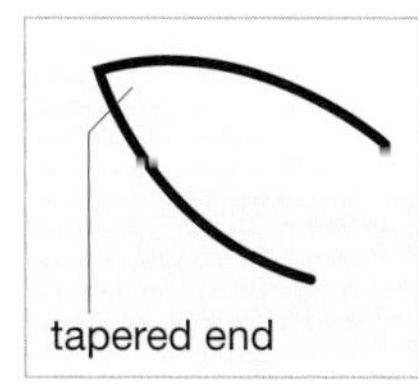

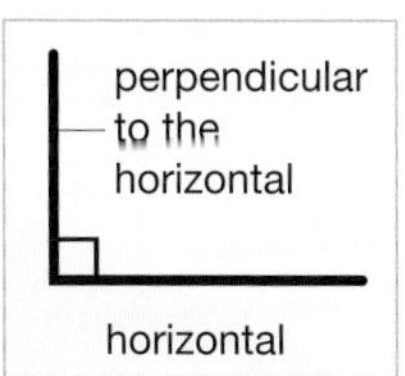

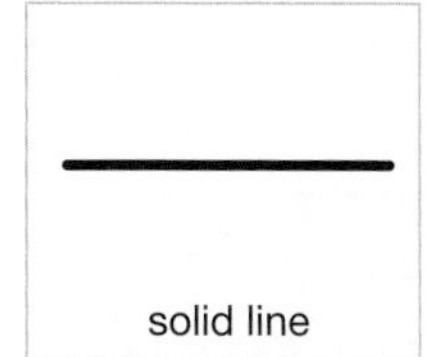

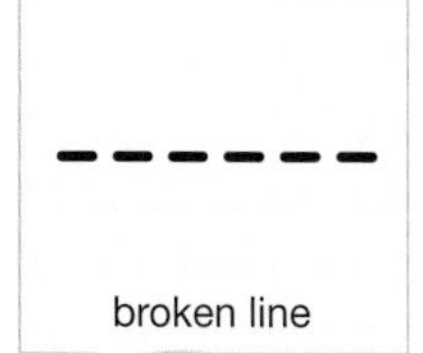

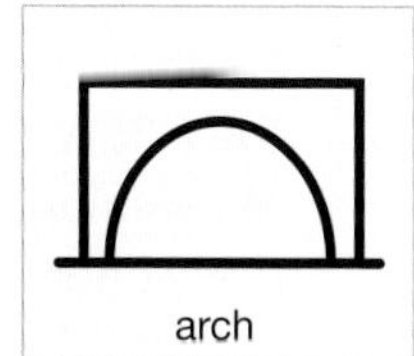

If you look at the photo and sketches of the Administrative Building, you can see an interesting architectural feature on the (1) __________ end of the building. The hatched area on the plan and the elevation represents a square (2) __________, which consists of a simple (3) __________ beam which is (4) __________ to a vertical column.

Turning now to the photo and sketches of the Student Hostel, the most interesting feature is the cylindrical staircase. This is shown as a (5) __________ circular shape on the (6) __________, and a rectangular shape on the (7) __________. Here, the (8) __________ line shows the staircase structure, and the (9) __________ line indicates the roof of the building behind the staircase.

Speaking 6 Work in pairs. Describe buildings so that your partner can locate them. Listen to your partner's descriptions and locate the buildings.

Student A Turn to page 114 and follow the instructions.
Student B Turn to page 115 and follow the instructions.

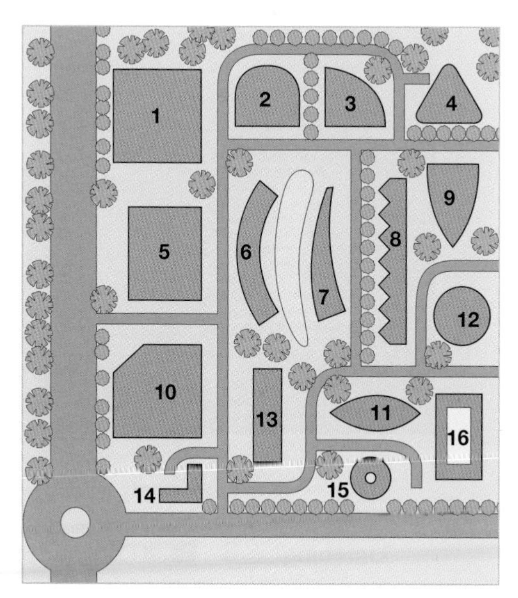

Scanning 7 Practise your speed reading. Look for the information you need on the SPEED SEARCH pages (116–117). Try to be first to answer these questions.

1 How heavy is the heaviest building that has been moved from one place to another?
2 Where is the smallest purpose-built cinema in operation?
3 Where is the deepest underwater mail box in the world? Why is it there?
4 How high is the highest concrete dam in the world.

Speaking 8 Talk to your partner. What world records have been mentioned in this book? What world records do you know about? Are there any world records in your industry? What records have been broken recently?

10 Disasters

1 Speculation

Start here

1 Work in small groups. What can cause a bridge to collapse? Make a list.

2 With your group discuss these photographs showing the I-35W bridge before and after its collapse. Discuss the possible causes of the collapse.

On the evening of 1 August 2007, the I-35W road bridge over the Mississippi River collapsed into the river, causing 13 deaths and 145 injuries. The government announced an immediate investigation into the causes of the disaster.

Listening

3 ▶ 33 After the collapse of the I-35W bridge, some technical experts called a radio phone-in show and suggested possible causes. Listen and tick the speculations that are mentioned.

1	compression	☐	6	impact	☐
2	bomb	☐	7	buckling	☐
3	thermal shock	☐	8	fracture	☐
4	corrosion	☐	9	metal fatigue	☐
5	wear	☐	10	tension	☐

Vocabulary

4 Match these phrases with words or phrases in 3 which have the same or similar meaning.

a) damage caused by continued loading ☐
b) temporary bending due to compression ☐
c) striking it with a hard force over a short period ☐
d) expanding or contracting suddenly because of extreme heat or cold ☐
e) disintegration caused by chemical reaction with oxygen and water ☐
f) breaking into two or more pieces ☐
g) removal of surface material by rubbing or friction ☐
h) a pulling or stretching force ☐

Language

We can use these language forms to make speculations.

Active	
	The speaker thinks that …
A girder **could / might / may have broken**.	… it's *possible* that it happened.
The girder **might / may not have broken**.	… it's *possible* that it *didn't* happen.
One of the bearings **must have corroded**.	… it's *certain* that it happened.
A bomb **can't / couldn't have caused** the damage.	… it's *impossible* that it happened.

Passive				
The collapse	could / might (not) / may (not) must can't / couldn't	have been	caused triggered	by buckling. by corrosion. by a bomb.

5 Mark these statements from the phone-in P (*possible*), C (*certain*) or I (*impossible*).

1 The collapse can't have been caused by a bomb. ☐
2 I think that one or more of the girders might have buckled. ☐
3 Well, I reckon the collapse could have been due to metal fatigue. ☐
4 My own view is that one of the bearings must have corroded and rusted away. ☐
5 Well, I think the collapse could have been caused by thermal shock. ☐
6 I think the collapse must have been caused by wear. ☐

6 These statements were made after investigations into other disasters. Change them into speculations made before the investigations. Use the word(s) in brackets. Then write P, I or C after each sentence.

Example: *1 The Challenger disaster could have been caused by a faulty O-ring seal. (P) The seal might …*

1 The Challenger shuttle disaster was caused by a faulty O-ring seal. (*could*)
The seal broke away from a fuel tank and damaged it. (*might*)
2 The wing of the Columbia shuttle was damaged by an insulating tile. (*must*)
The tile fell off the nose cone at launch. (*may*)
3 The Warsaw radio mast collapsed because of human error. (*may*)
The cables securing the mast were not fastened correctly. (*might not*)
The mast bent and then snapped into two. (*must*)
4 The crash of the Air France Concorde was not caused by a fault in the plane itself. (*can't*)
One of its tyres was cut by a metal strip lying on the runway. (*must*)
Another aircraft dropped the strip on the runway some minutes before. (*could*)

Speaking

7 In pairs, talk about some dangerous or unusual events which have happened to you in your work or in your life. Speculate about the possible causes of the events.

2 Investigation

Start here

1 Match the words in the box with the labels 1–6.

truss girder gusset plate bearing pier deck

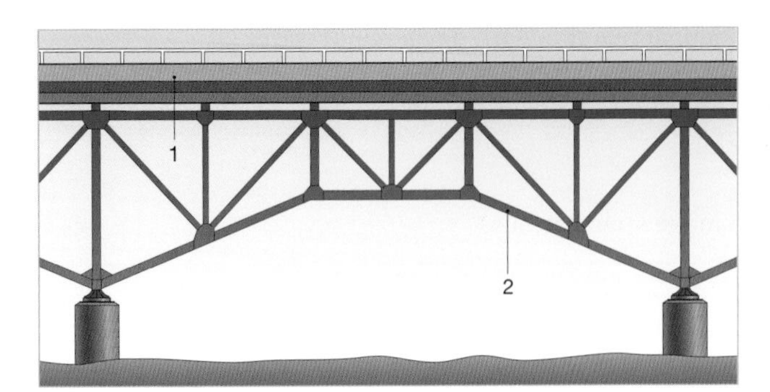

Scanning

2 Practise your speed reading. Look for the information you need somewhere on pages 78–79. Try to be first to answer these questions.

1 How many metres of the bridge fell into the river?
2 How many lanes were open to traffic on the day of the collapse?
3 How many vehicles were recovered from the water?
4 How heavy was the construction equipment on the bridge on the day of its collapse?

Listening

3 34 Listen to this interview between an investigator and the company that managed the collapsed bridge. Put a tick in the correct boxes.

ACTION	YES	NO
1 bridge inspected every year		✓
2 corrosion discovered in bearings		
3 corroded bearings repaired or replaced		
4 cracks found in girders		
5 cracks in girder drilled to prevent spreading		
6 support struts added to cracked girders		
7 signs of metal fatigue discovered		
8 bridge closed after metal fatigue discovered		
9 steel reinforcement carried out on bridge		
10 design error in gusset plates discovered before collapse		
11 undersized gusset plates replaced with larger ones		

4 Complete these sentences from the interview, using the words in the box. Check your answers in the audioscript on page 125.

been had have wouldn't

1 Your company should ______ inspected the bridge annually. Did a competent employee of your company carry out an annual inspection?
2 The bridge should ______ ______ inspected in 2007. What did previous inspection reports say?
3 The bearings shouldn't ______ ______ left on that bridge.
4 If your company ______ replaced the bearings, maybe the bridge ______ ______ collapsed.

Language We can use *should* with the perfect infinitive to criticise a past action that was wrongly taken, or not taken.

Active: Your company *should / shouldn't have replaced* the bearings.
Passive: The bearings *should / shouldn't have been replaced.*

5 Look at the actions you marked NO in 3. Make sentences to criticise the fact that the actions were not done.

Example: *1 (bridge not inspected every year)* ➔ *The bridge should have been inspected every year.*

Language We can use the third conditional to speculate about unreal events or situations in the past.

If you had replaced the bearings,	the bridge would not have collapsed.*
If the bearings had been replaced,	the bridge would be standing now.**

* unreal result in the past ** unreal result in the present

6 Speculate about unreal situations in the past.

Example: *1 If the main column had not buckled, the building would not have collapsed.*

1 The main column buckled and then the building collapsed.
2 The plane's fuel tank fractured and then the fuel exploded.
3 A ship crashed into the bridge pier and now the pier is cracked.
4 Friction wore down the brake pads and so the brakes do not work now.
5 Tensile forces stretched the cables and as a result the cables snapped.
6 Compressive forces pressed down on the columns, and as a result they are now fractured in three places.

Task **7** Work in groups of four. Discuss the collapse of the hotel walkways. What caused it to happen? What should people have done to prevent it?

Student A: Study the information on page 110.
Student B: Study the information on page 112.
Student C: Study the information on page 114.
Student D: Study the information on page 115.

8 Briefly explain your group's conclusions to the class. Say what happened, who or what caused it to happen and what should have happened?

3 Reports

Start here 1 Work in pairs. Match the section headings with their explanations. Do you think these are the best section headings for a report of an investigation? Do you agree with the order of the sections? Do you write reports in a different way? Discuss the best format for the reports you have to write.

Writing an investigative report	
Report sections	in which you …
1 Abstract 2 Introduction 3 Background 4 Method 5 Findings 6 Conclusions 7 Recommendations 8 Attachments	a) give history to explain why the investigation was necessary b) present the evidence or data that you discovered c) summarise the whole report d) give your overall opinion based on all the evidence e) say what action others should take as a result of the report f) explain briefly your purpose in writing the report g) provide additional tables, diagrams, reference and documents h) show how you carried out the investigation

Abstract can also be called *Summary*. *Method* can also be called *Procedure*.

Reading 2 Write the correct numbered headings from 1 for each section of the investigation report of the I-35W bridge collapse.

8 Attachments

Fig. 1: Diagram showing the gusset plate joining two girders

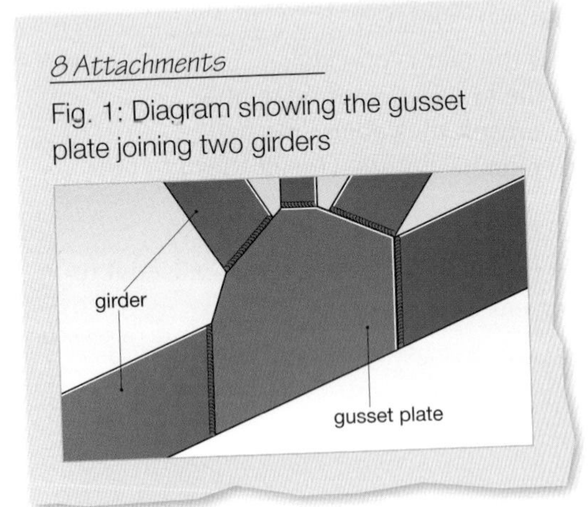

____

After studying all the evidence, the investigating team has decided that the collapse of the bridge must have been caused by the weakness of the gusset plates which connected a group of girders together. As indicated earlier, these plates were too thin to withstand (1) the 20% increase of load over the years, or (2) the additional traffic and construction loads on the day of the collapse. It is clear that if the gusset plates had been of the correct thickness, the increased loads on the bridge would probably not have caused the collapse.

____

In order to carry out a full investigation into the causes of the collapse, the investigating team used a range of procedures. Footage from security CCTV cameras near the bridge was studied. Wreckage from the bridge was recovered and examined. Photographs and documents of previous inspections of the bridge were looked at. Inspectors and construction and maintenance staff were interviewed.

____

1. Non-destructive tests should be carried out on all gusset plates in bridges.
2. A system should be introduced for detecting bridge design errors at an early stage.
3. Training courses for bridge inspectors should be updated.
4. The bridge inspection manual should be revised to include data on gusset plates.

____

The purpose of this document is to present the results of the investigation into the collapse of the I-35W bridge on August 1 2007.

During the investigation, the following evidence was discovered:

1. Photographs from inspections before the collapse show that a number of bearings and gusset plates were badly corroded.
2. The gusset plates which connected the girders at the U10 nodes showed signs of cracking, fracturing and buckling.
3. Some of the gusset plates were found to be too thin, and of inadequate strength to support the loads on the bridge. The plates were 13 mm thick, but they should have been several millimetres thicker.
4. Annual inspections before the collapse did not reveal the problems with the gusset plates.
5. In the years before the collapse, 51 mm of concrete were added to the road surface, increasing the load by 20%.
6. On the day of the collapse, there were additional loads on the bridge due to construction work. There were 261,000 kg of construction equipment on the bridge. Four lanes were closed to traffic so that there was more traffic concentrated in the four open lanes.

This report presents the results of the investigation into the collapse of the I-35W bridge. It describes the method of investigation, including CCTV footage, wreckage, photographs, inspection records and interviews with staff. The main findings are (1) buckling and fracturing of undersized gusset plates, (2) corrosion of some components, (3) excessive load on the bridge on the day of the collapse, and (4) inadequate inspections over the years. The investigation concludes that the main cause of the collapse must have been the undersized gusset plates which failed under excessive loads on the day of the collapse. The report recommends (1) a system for detecting design errors, (2) non-destructive testing of gusset plates, and (3) updating of inspector training courses and manuals.

At about 6:05 pm on August 1, 2007, the eight-lane, 581-metre-long I-35W highway bridge over the Mississippi River in Minneapolis, Minnesota, experienced a catastrophic failure in the main span of the deck truss. As a result, about 300 metres of the deck truss collapsed, with approximately 140 metres of the main span falling 33 metres into the 4.5-metre-deep river. A total of 111 vehicles were on the portion of the bridge that collapsed. Of these, 17 were recovered from the water. As a result of the bridge collapse, 13 people died, and 145 people were injured. An immediate investigation into the causes of the collapse was ordered.

3 Answer these questions about the report.

1. According to the conclusions, which parts of the bridge failed?
2. What design error caused these parts to fail?
3. What happened on the day of the collapse which added to the load on the bridge?
4. What physical evidence of the collapse did the investigators examine?

Task **4** In pairs, decide which sections (*Introduction*, *Findings*, etc.) of an investigative report would contain these sentences. Then discuss the verb form used in each case, and why.

1. The accident must have been caused by a fault in the railway signals.
2. New safety equipment should be supplied to all staff working on the rig.
3. The ship should have been inspected for cracks and fractures every year.
4. Twenty-six near-miss incidents were reported to central air traffic control.
5. This report gives an account of the investigation into the recent fire.

Writing **5** Work with the same group you worked with on page 77. Produce your group's report on the investigation into the Hyatt Regency disaster. Use the same format as the one used in 1 and 2 above.

Each group member should write one or more sections. Appoint a group leader. The group leader should check that everyone has a roughly equal amount of writing to do.

Review Unit E

1 Write six sentences about any two of the three smartphones. Use comparative adjectives. Modify the comparisons in a ***general*** way (see the language box on page 68).

Examples: 1 *The Nokia is slightly longer than the iPhone.* 3 *The iPhone is much thinner than the Palm Pre.*

	Apple iPhone 3G S	Nokia N97	Palm Pre
1 Length	115.5 mm	117.2 mm	100.5 mm
2 Width	62.1 mm	55.3 mm	59.5 mm
3 Thickness	12.3 mm	15.9 mm	16.95 mm
4 Weight	135 g	150 g	135 g
5 Camera (resolution)	3 Mp	5 Mp	3 Mp
6 Display	89 mm touchscreen	89 mm	78 mm
7 Talk time (max)	720 min	400 min	180 min
8 Capacity	16 / 32 GB	32 GB	8 GB
9 PAYG price*	16 GB £440 / 32 GB £538	£499	[to be announced]

*price announced at launch of the PAYG (Pay As You Go) model

2 Make comparisons from the table in 2 using the notes below. Modify the comparisons in a ***specific*** way (see the language box on page 68). Use the words in the box.

approximately roughly about virtually almost exactly

Example: 1 *The capacity of the 32 GB iPhone is exactly four times as great as (or greater than) the capacity of the Palm Pre.*

1 capacity / four times / great
2 display / ten millimetres / small
3 Apple / 25% / thin / Palm Pre
4 Nokia / twice / pixels / Apple
5 talk time / four times / long
6 32 GB model / £40 / expensive

3 Make eight statements about the devices, using superlative adjectives. Use ***easily*** or ***by far*** where appropriate.

Example: *The iPhone has easily the longest talk time of the three.*

4 Change these sentences to give the same meaning by reversing the order of the items compared.

Example: 1 *(begin:) Capital Gate is …*

1 Hearst Tower is much taller than Capital Gate.
2 Capital Gate has a slightly larger floor area than the Gherkin.
3 The Gherkin is a little less tall than Hearst Tower.
4 A normal tower consumes twice as much energy as the Gherkin.
5 The Bird's Nest in Beijing is about five times as heavy as Capital Gate.
6 A normal building uses about 25% more steel than Hearst Tower.

(Note: from 80 to 100 is a *25% increase*. From 100 to 80 is a *20% decrease*.)

5 Match each object A–D with its description 1–4.

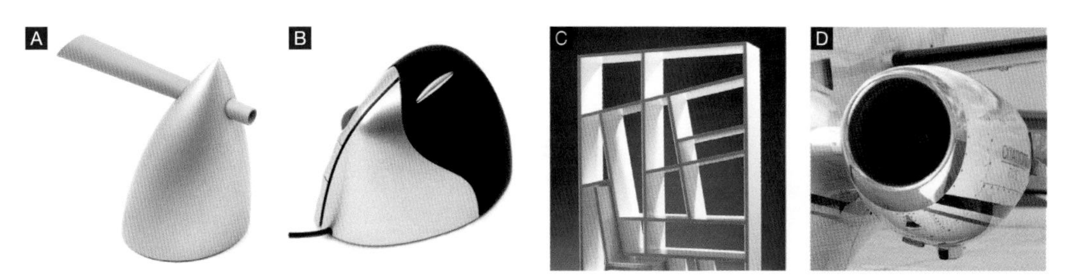

1. This structure has a rectangular frame which contains many flat surfaces. The surfaces are horizontal, vertical and at different angles.
2. This item is roughly hemi-spherical in shape, with one side slightly tapered. Its base forms a triangular shape with rounded corners.
3. This item is approximately cylindrical in shape, but slightly tapered towards one end.
4. This item is roughly conical in shape and bulges slightly around its middle. Near the top, another item is inserted through it. This second item is hollow, roughly cylindrical, hollow and slightly tapered.

6 Describe the shape of these objects, devices or structures in a similar way to the descriptions in 5. Use words and phrases from the box.

rounded curved tapered flattened rectangular triangular circular semi-circular cylindrical conical elliptical L-shaped star-shaped shell-shaped zigzag at an angle

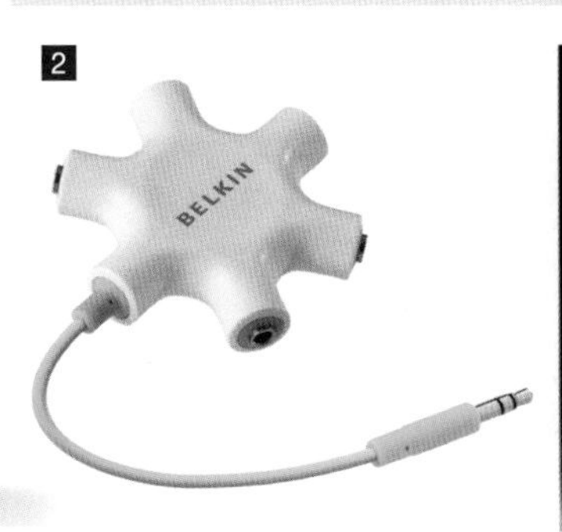

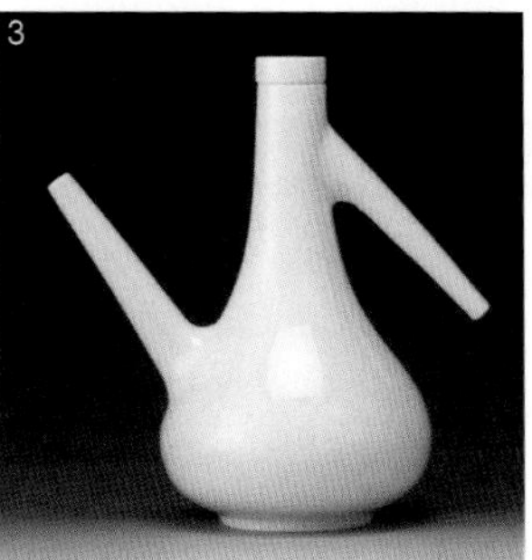

7 Complete the text with the letters A–H from the technical drawing.

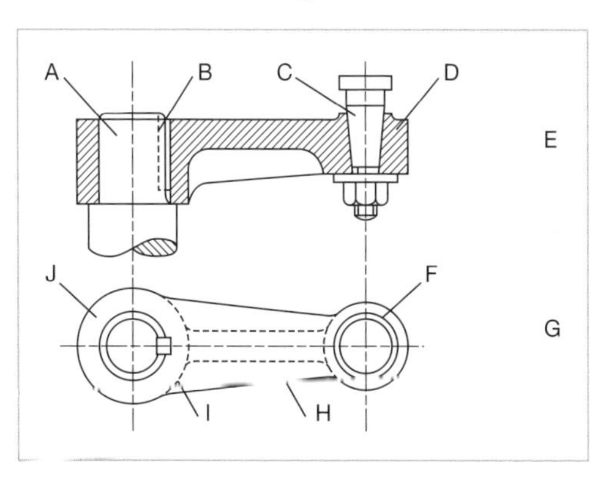

First let's look at the elevation view (1) ______ of the component. Here you can see a nut, and above this you can see a tapered cylindrical shaft (2) ______. Now I'd like you to study the hatched area (3) ______ adjacent to the nut and shaft. On the opposite side of the component, still looking in the elevation view, you can see a cylindrical shaft which is not tapered (4) ______. Now study the broken line (5) ______ which runs along one side of this shaft.

Now let's turn to the plan view (6) ______ of the component, where you can see two ring-shaped (or doughnut-shaped) structures. Look at the larger one (7) ______. Now look at the smaller one (8) ______. The two ring or doughnut shapes are connected by two solid lines, which are inclined at an angle to the horizontal. Look at the lower one (9) ______. Finally, let's look at the curved broken line (10) ______ which is adjacent to the lower solid inclined line.

8 Speculate about what *might have happened*, what *must have happened* and what *can't have happened* in these situations. Use the verbs in the box.

must have   can't have   may have   might have   could have

9 Look at this summary and make statements about what should or should not have been done. Use the passive without an agent.

Example: 1 *The automatic locks on the fire doors should not have been disabled.*

Summary of mistakes made in the weeks and months leading up to the fire:

1. Staff disabled the automatic locks on the fire doors.
2. Contractors did not inspect the fire sprinklers every month.
3. Staff took away the fire extinguishers for servicing all at the same time.
4. Managers did not carry out fire drills on a regular basis.
5. Workers demolished two of the fire escapes.
6. Safety staff did not test the fire alarm system every two months.

10 Imagine that it is 1915. Make recommendations for the future from these conclusions, using your own ideas. Use *should* with the passive infinitive.

Possible answer: 1 *In future, the steel used in ship hulls should not be made of …*

REPORT ON THE SINKING OF THE *TITANIC*

CONCLUSIONS

1 The type of steel used in the hull was unsuitable for very cold weather. This type of steel becomes very brittle in cold temperatures. As a result, when the iceberg struck the ship, the hull buckled and punctured easily.

2 The rivets of the hull were of sub-standard quality. The metal of the rivets contained too much slag, which made the rivets break easily.

3 The lookouts did not see the iceberg in time, because they were not using binoculars. The lookout gave the alarm only a short time before the ship struck the iceberg, so the ship was unable to turn out of danger in time.

4 There were not enough lifeboats to hold all of the crew and passengers. There were enough lifeboats for 1,178 people, one third of the *Titanic's* total capacity of 3,547 people.

11 Complete these sentences with the nouns or noun phrases in the box.

tension metal fatigue corrosion fracture impact thermal shock wear

1 If you continue to overload the bridge deck day every day for years, after a while it will fail due to *metal fatigue* .
2 The railway line buckled due to ___________. This was because it expanded and cooled every day in the desert climate.
3 The ___________ of the hammer striking the concrete slab made it ___________ into two pieces.
4 The constant friction of the disc on the brake pad made the pad become thinner due to ___________.
5 The weight of the bridge deck pulling down on the cables was too much. After a while, the ___________ made them stretch and then break.
6 The girder has been destroyed by the action of oxygen, water and chemicals on the girder, i.e. the cause of the accident was ___________.

12 Complete these sentences using the third conditional. Use the correct form of the verbs in brackets. Each gap contains two, three or four words, including at least one word from the box.

had have would be been

1 If the concrete *had been mixed* (mix) to the correct proportions, the bridge pier *would* ___________ (not / collapse) in the floods, and the bridge ___________ (still / stand) now.
2 If the lookouts on the *Titanic* ___________ (use) binoculars, or if radar ___________ (invent) in the 1910s, perhaps the ship ___________ (not / collide) with the iceberg.
3 If debris ___________ (not / fall) onto the runway, the aircraft ___________ (not / strike) it. As a result, the aircraft ___________ (not / come) off the runway, and all the passengers ___________ (be) alive now.
4 If the shuttle *Columbia* ___________ (equip) with a launch abort system, the launch ___________ (stop) and the crew capsule ___________ (eject) from the shuttle.

13 Make statements using the third conditional.

Example: 1 *If the engineer had not tripped the wrong switch, the electrical system would not have broken down last night.*

1 The electrical system broke down in the city last night because an engineer tripped the wrong switch.
2 The earthquake destroyed over 20% of the town because the buildings were not built with earthquake-proof foundations.
3 The 27 accidents took place because the police took down all the speed cameras from this section of the road.
4 The pilot of the aircraft did not take evasive action because the cloud prevented him from seeing the other plane.

Projects **14** Research one of the following and then present your information in a short talk to the class or to a group.

1 A serious incident or accident that happened in your industry. Write the main points under the headings of an investigative report.
2 Failures of design, construction, engineering or marketing. List (a) what went wrong (b) why they went wrong (c) what should have happened (d) what would have changed if different decisions had been made.

11 Materials

1 Equipment

Start here

1 Work in small groups and choose a sport. Make a list of the main equipment used, then draw up a chart like this.

Sport	Equipment	Material	Type	Properties
climbing	rope	nylon	synthetic fibre	tensile strength, elasticity
skiing	ski pole	graphite	composite	lightness, rigidity

2 With your group, explain your chart to the class. Tell them what materials are used and why they are used in the sport you have chosen.

Listening

3 35 Listen to this phone call and answer the questions.

1 What is the purpose of the call?
2 What happened before the call?
3 What will happen as a result of the call?

Reading

4 Read this letter and say which word or phrase in the box best describes it.

letter of thanks invitation invoice presentation proposal attachment covering letter personal letter application

Dear Ramón,

Thank you very much for inviting me to make a proposal to supply your team with DesignerSport football boots.

At the presentation I gave at your HQ last week, I demonstrated the boots and explained how I think we can help your team to win the next Cup Final. In our phone call the following day, you kindly invited me to send you a proposal.

As I explained, our boot is designed to give protection, lightness and comfort. After the famous foot injuries of Beckham and Rooney in the England team, football managers want to provide more protection for their players.

Our boot combines the lightness of carbon fibre with the strength of aramid fibre. Details of all the materials used in the boot can be found in the attachment to this letter or, alternatively, by clicking on www.footieboots.com/audio.

The boot gives the player torsional stability. The studs are injection moulded on to the bottom of the boot, allowing the player to accelerate across the field.

My company proposes to supply these boots at the unit price (per pair) of €49.50. Package and delivery is free of charge, and delivery dates are a maximum of two weeks after placing the order. This offer is open for 28 days from the date of this letter.

I look forward to hearing from you with a firm order in due course.

Best wishes,

Albert Weston

Albert Weston
Manager, DesignerSport

5 Answer these questions about the letter.

1 What two things did Albert do before he wrote the letter?
2 What action does Albert want Ramón to take next?
3 How does Albert try to convince Ramón that aramid fibre is needed in a boot?
4 Which sentence explains that the boot helps the player to twist his body easily without slipping or falling?
5 Can the studs to be unscrewed or removed easily from the boot?
6 What will happen if Ramón waits for six weeks to make his order?

Listening

6 36 Listen to www.footieboots.com/audio and fill in the gaps in the attachment to the letter with the words in the box.

soft strong in tension lightweight impact absorbent
elastic flexible tough impact resistant

Materials used in the DesignerSport football boot

Part of shoe	Material	Key properties
Upper (top)	carbon fibre	(1) ______ (2) ______
	aramid fibre	(3) ______ (4) ______
Padding (inside)	polyurethane foam	(5) ______ (6) ______
Sole (bottom)	thermoplastic polyurethane (TPU)	(7) ______ (8) ______

carbon fibre aramid fibre polyurethane foam
thermoplastic polyurethane TPU impact absorbent impact resistant 37

Language

We can describe the properties of materials in a variety of ways.

present simple active	this material **resists** heat; it **weighs** very little
can / can't	you **can stretch** it; it **can / can't be bent** easily
active with passive meaning	*it **bends easily** (= it can be bent easily); it **burns easily** (= it can be burnt)*
without *-ing*	*it absorbs a blow **without bending** (= but it doesn't bend)*

7 Complete these sentences, using the information in 6. Use the correct form of the verbs in the box. Some verbs are used more than once.

transfer stretch return weigh break reduce bend

1 Carbon fibre doesn't ______ very much, and it can be ______ easily.
2 Aramid fibre can't be ______ easily, and it can withstand a heavy blow without ______.
3 Polyurethane foam is soft, and so it ______ the effect of a violent blow without ______ the impact to the wearer.
4 TPU doesn't ______ easily, but it can be ______ and then it can ______ to its original shape.

2 Properties (1)

Start here

1 What properties do these materials have? Match the words in the box to the photos 1–4.

malleability tensile strength flexibility compressive strength 38

2 Discuss in pairs. What properties must these items have?

1 a firefighter's protective jacket
2 a concrete beam in a skyscraper
2 a scuba diver's watch
4 the metal for an electric cable

Scanning

3 Practise your speed reading. Look for the information you need on the SPEED SEARCH pages (116–117). Try to be first to answer these questions.

1 How many plastics have the property of impact resistance?
2 Which two plastics are used in making arrows?
3 How many plastics have the property of durability?
4 Which two plastics are used in clothing?

Vocabulary

4 Match the adjectives and adjectival phrases 1–6 with their meanings a–f.

1 malleable
2 heat tolerant
3 ductile
4 absorbent
5 strong in torsion
6 strong in compression

a) able to take in liquids; able to reduce the effect of impact
b) able to be pulled into a longer, thinner shape without breaking
c) capable of resisting a twisting force
d) able to withstand a heavy force pressing down on it
e) capable of being hammered or rolled into a new shape
f) capable of withstanding heat without being affected

5 Change these nouns into adjectives.

1 malleability 2 non-flammability 3 tolerance 4 ductility 5 durability 6 absorbency

6 Rewrite these sentences to give the same meaning.

Note: *shear* is both an adjective and a noun.

1 Steel cable has *good tensile strength*. ➔ Steel cable is <u>very strong in tension.</u>
2 Concrete has *high compressive strength*. ➔ Concrete is ...
3 Nylon is *extremely strong in tension*. ➔ Nylon has ...
4 This metal has *excellent torsional strength*. This metal is ...
5 This steel is *very strong in compression*. ➔ This steel has ...
6 Kevlar has *very good shear strength*. ➔ Kevlar is ...

Language We often describe the properties of *resistance* and *tolerance* in these ways.

verb	This material **resists** water very well.
adjectival phrase	This material **is** highly **water resistant**.
hyphenated adjective	This **is** an extremely **water-resistant** material.
noun phrase	This material **has** very good **resistance to water**.
noun phrase	This material **has** excellent **water resistance**.

7 Make a chart like the one in Language for each of these sentences.

1 This plastic tolerates heat extremely well.
2 This plastic resists chemicals very well. (Be careful with the *-s* on *chemicals.*)

8 Complete this text.

KEVLAR® is increasingly popular with manufacturers of motorcycle components and protective clothing. This is because of its (1) ___________ (rigid / rigidity), its (2) ___________ (tolerant / tolerance) to heat and its high impact (3) ___________ (resistant / resistance).

Hiking boot manufacturers use Kevlar® in uppers, soles and laces because it is more (4) ___________ (durable / durability) than other fibres. Kevlar® also appears in hiking clothes because it is highly (5) ___________ (resistant / resistance) to tears and abrasion from branches and rocks.

Tennis, squash and badminton racquet strings are often made of Kevlar® because it is highly (6) ___________ (stretch resistant / stretch resistance). Kevlar® frames (7) ___________ (resistant / resistance / resist) cracking and fracturing. They are also incredibly (8) ___________ (lightness / lightweight) and have extreme (9) ___________ (rigid / rigidity).

Kevlar® makes high-performance skis and ski boots lighter, more (10) ___________ (rigid / rigidity) and more responsive, and (11) ___________ (absorbs / absorbent / absorbency) vibration and impact very well. Speed, (12) ___________ (stable / stability) and good turning ability are the qualities expected for high performance skis.

Writing 9 Make a chart like this about the main materials used in your technical field, and their properties.

Example: *Technical field: Construction*

Application	Material	Properties
beams and columns	reinforced concrete	rigidity; compressive strength

10 Write a short description based on the table you produced in 9.

Example: *In the construction industry, beams and columns are often made of reinforced concrete. This material is used because of its rigidity and compressive strength (or because it is rigid and strong in compression).*

3 Properties (2)

Start here

1 Work in small groups. Imagine you are managers and trainers of your country's Olympic team. What decisions would you take about training, clothing, equipment and other matters to improve your team's performance?

Listening

2 39 Listen to this meeting and answer the questions.

1 What two important issues are discussed in the meeting?
2 What (a) land-based sports and (b) water-based sports are discussed?

3 Listen to the meeting again and complete the minutes. Next to ***Reason(s)***, write in the property or properties of the equipment or material to be investigated.

> The executive Olympic team committee of managers and trainers met on 4th August and decided to investigate the following ideas for the next-but-one Olympics:
>
> **A** Flite shoes. Reasons: 1 ____________ 2 ____________
> **B** Marathonite shoes. Reasons: 3 ____________ 4 ____________
> **C** SpeedShark material. Reason: 5 ____________
> **D** Sensors attached to oars. Reason: *able to provide* 6 ____________
> **E** Doppler lidar system. Reason: *able to provide* 7 ____________

4 Complete these phrases used in the meeting by people making suggestions. Then listen again and check your answers.

1 Why ______ ______ try the new Flite shoes?
2 ____________ try using the newest Marathonites.
3 Can I ____________ a suggestion? We ____________ look at the new SpeedShark swimsuit.
4 I ____________ suggest ____________ we need to invest more in sensors.
5 How ____________ starting with the rowing team?
6 ____________ look into all these suggestions and make a full report.

Language Ways of making a suggestion:

Why don't we do …? Let's try doing … I would suggest that we do …	Let's do … What / How about doing …? We could do …

Speaking 5 Work in pairs or small groups. Have a brainstorming session about how to improve your college/your work place/a sporting team/an everyday device (such as a mobile phone).

Language

<table>
<tr><td rowspan="5">This material</td><td colspan="2">can(not)</td><td></td><td rowspan="3">resist impact.
withstand pressure.</td></tr>
<tr><td>is</td><td>(un)able</td><td rowspan="2">to</td></tr>
<tr><td>has</td><td>the ability/ capability/ capacity</td></tr>
<tr><td>is</td><td>(in)capable</td><td rowspan="2">of</td><td rowspan="2">resisting impact.
withstanding pressure.</td></tr>
<tr><td>has</td><td>the capability</td></tr>
</table>

6 Rewrite these sentences to give the same meaning. Use the words in brackets. Do not use the words in italics.

1. Kevlar is used in bulletproof vests because it has the *ability* to resist severe impact. (capable)
2. Polypropylene is used inside crash helmets because it has the *capacity* to absorb impact and to soften a blow to the head. (capability)
3. Nylon is commonly used in waterproof jackets because it is *capable* of withstanding water and preventing it from passing through. (capacity)
4. Wood is rarely used nowadays for making a boat hull because it is *unable* to stop rocks from cutting into the hull. (incapable)

Vocabulary 7 Build up a list of similar words useful in your technical field. Check them in a dictionary or online search engine before you write them in the box.

proof	**resistant**
fireproof, bulletproof, ovenproof (dish), childproof (locks)	*fire resistant, water resistant, stain resistant, shock resistant, corrosion resistant*

Note use of the hyphen: *The door is fire resistant. It's a fire-resistant door.*

A *fireproof / fire-resistant* door = a door which resists fire; a door which does not allow fire to pass through it; a door which stops fires from spreading.

Note: *proof* is generally stronger than *resistant*; for example, a waterproof jacket is probably guaranteed to keep out all water, but a water-resistant jacket may not be.

Task 8 Work in groups of four. You are some of the Olympic team of managers and trainers who held the meeting in 2. You have been called to a second meeting and you have to make some new decisions using the information below. Have the meeting and reach your decisions.

The government has informed the Olympic national team that it must reduce its budget for equipment and training by 60%, because of the economic downturn. Appoint a chairperson (Student D) for the meeting.

Student A: Turn to page 109.
Student B: Turn to page 111.
Student C: Turn to page 112.
Student D: Turn to page 114.

Writing 9 You are the chairperson of the Olympic team who held the meeting in 8. With your group, write a memo from your team to the head of the national Olympic Committee explaining the decisions your team has made.

12 Opportunities

1 Threats

Start here

1 Discuss these graphs in pairs.

1. What predictions does each graph show?
2. Which predictions are the *best-case scenario* and which ones are the *worst-case scenario*?
3. Which predictions do you think are most likely? Why?

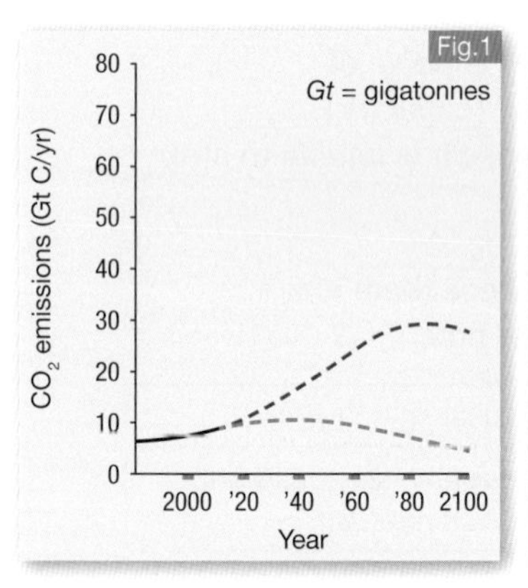

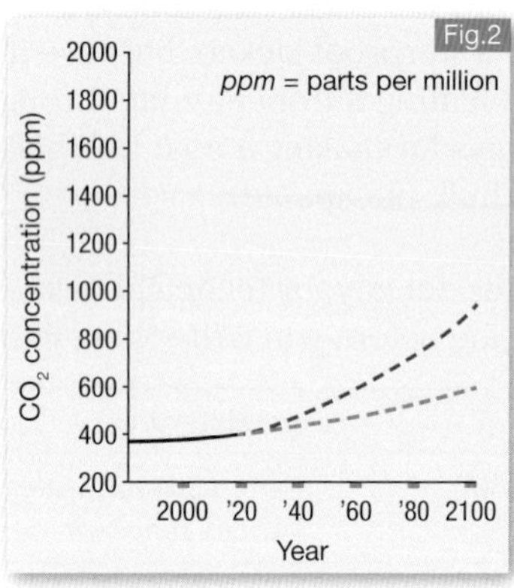

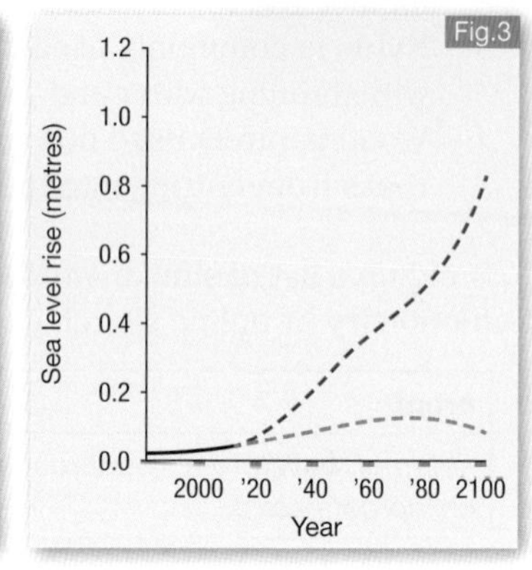

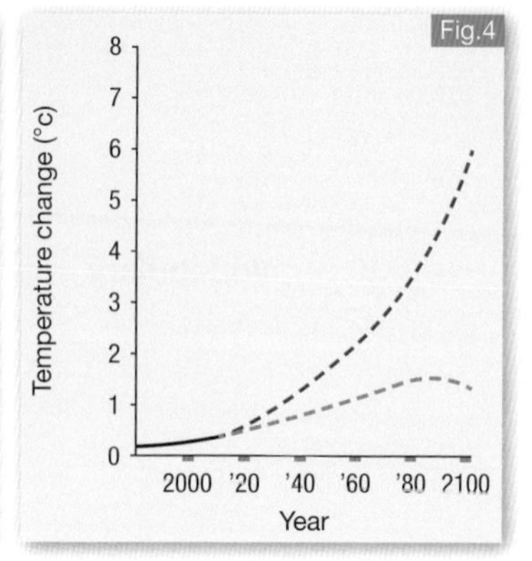

Listening

2 A message has arrived for the people of today from people in the year 2060. Do you think the news is good or bad?

3 

Listen to the first part of the message from the future. What actually happens to the four variables in the graphs by the year 2060?

4 Listen again and make a note of what has happened to the following locations.

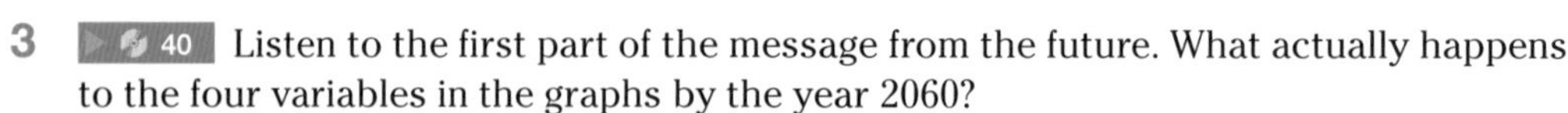

Example: *Fires have burnt down huge areas of forest.*

5 The next part of the message is partly lost through radio interference. It talks about present-day society's mistakes. With your partner, fill in as many gaps as you can before you listen. Don't worry if you can't fill in all the gaps.

Your society should have (1) ____________ your (2) ____________ of oil and other fossil fuels. You should (3) ____________ invested more in renewable energy. Your governments (4) ____________ have encouraged cheap air flights; instead, they (5) ____________ (6) ____________ put higher taxes on air fuel to (7) ____________ the cost of air travel. Everyone should (8) ____________ (9) ____________ their own energy in their homes. They should have (10) ____________ wind turbines and solar panels on their houses. Why (11) ____________ your society and governments do these things? If you had (12) ____________ out these actions, the world's temperature probably would not (13) ____________ (14) ____________ by eight degrees Celsius. If your government (15) ____________ (16) ____________ better decisions, the sea level (17) ____________ probably not (18) ____________ (19) ____________ by one point two metres, and low-lying areas would not have been (20) ____________ .

6 41 Listen to the next part of the message. Check your answers in 5 and try to fill in all the gaps. (Note: you will not be able to hear all of the words because of the radio interference, so you should guess what the words are.)

7 With another pair (or the class), discuss the answers you had to guess. Explain why you think they are correct.

Speaking

8 Explain the difference between the situation in 2060 and (a) now, (b) the worst-case predictions in the graphs in 1, and (c) the best-case predictions. Go round the class giving different statements.

Example: *The actual increase in CO_2 emissions by 2060 will be more than 160% higher than the worst-case predictions today.*

Listening

9 42 Listen and complete this extract of a scientist's summary.

By 2060, CO_2 emissions (1) ________________ to 80 gigatonnes per year. CO_2 concentrations in the atmosphere (2) ________________ to 2000 ppm. The world's temperature (3) ________________ up by eight degrees from today's levels.

Language The future perfect

active	By 2060,	the sea level	**will have risen** 1.2 metres.
passive		many low-lying countries	**will have been flooded**.

In the above examples, the rise in sea level and the flooding happen before 2060.

10 Complete the summary about what will have happened by the year 2060. Use the information you heard in 3 and 4.

1 Fires will have ________________.
2 Most of the world's forests ________________.
3 The Arctic ice cap ________________.
4 Mountain glaciers ________________.
5 Tropical cyclones ________________.
6 Water in many villages ________________.

Speaking

11 Work in pairs. Discuss a company, an industry or technical field you both know something about. Complete a SWOT analysis chart with notes about the *strengths* and *weaknesses* of the industry, and the *opportunities* and *threats* facing it over the next 10–15 years.

12 In pairs, discuss and make a note of actions which should be taken soon to (a) avoid the threats and (b) increase the opportunities.

13 Tell the class about your predictions for the industry by (approximately) 2025. Explain what will have happened if the actions you noted in 12 have been (or have not been) taken.

Example: *(If we don't make use of nanotechnology), (by 2025) our automotive industry will have fallen behind its competitors in fuel efficiency.*

2 Innovation

Start here **1** Work in small groups. Look at the information about the Greenbird. How can it travel so fast and yet remain so stable? Discuss this and make notes.

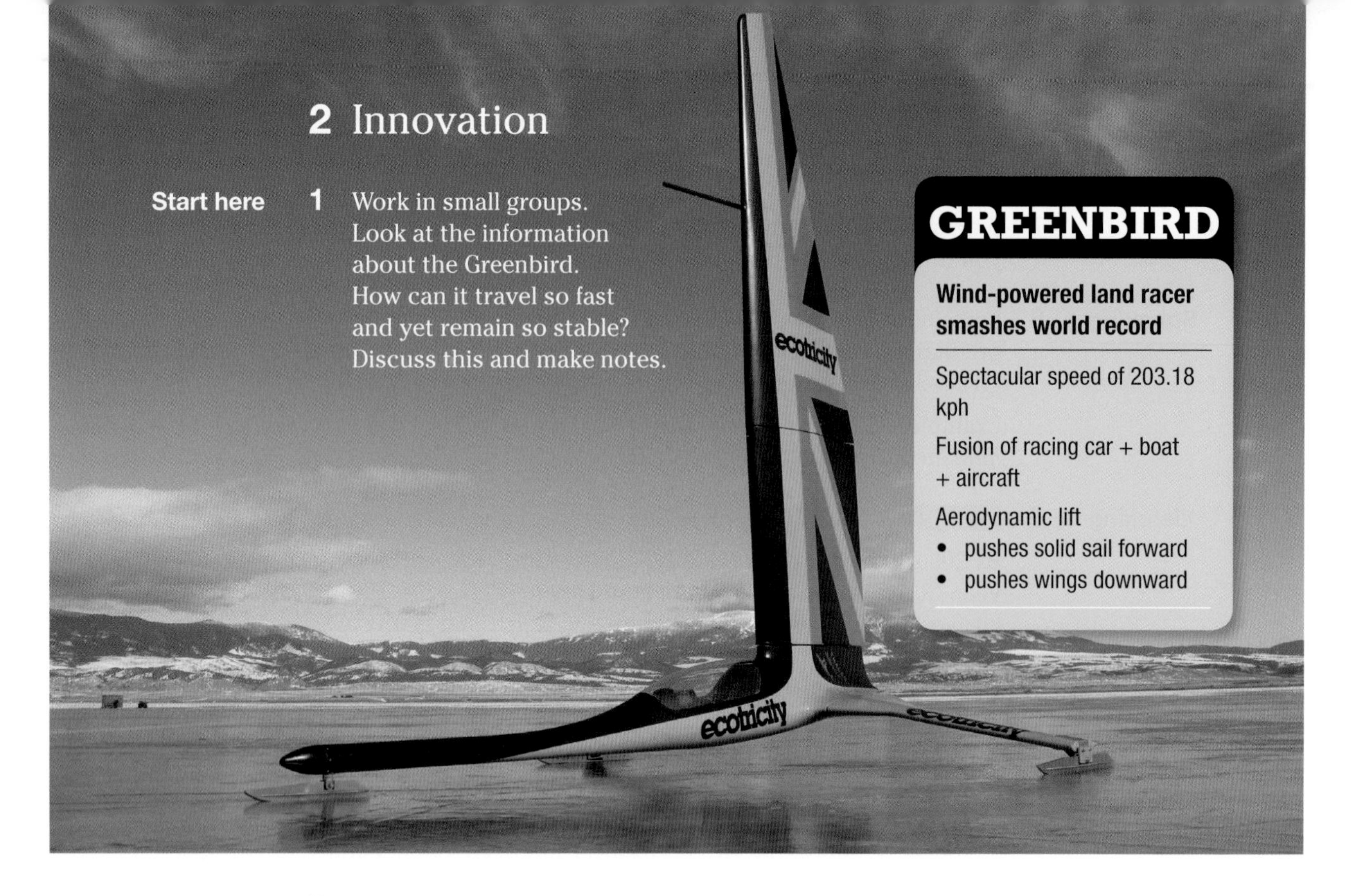

GREENBIRD

Wind-powered land racer smashes world record

Spectacular speed of 203.18 kph

Fusion of racing car + boat + aircraft

Aerodynamic lift
- pushes solid sail forward
- pushes wings downward

Reading **2** Read the article and compare the information with the notes you made in 1.

THE GREENBIRD relies on a solid sail, which is very similar to an aircraft wing. Just as airflow over a horizontal wing pushes an aircraft up, the flow of air over the Greenbird's vertical wing propels the vehicle forwards. This force enables the craft to travel at a staggering speed of five times the speed of the wind.

In addition, the Greenbird uses Formula One racing car technology to achieve exceptional stability. Made from carbon composites, the vehicle can withstand a massive sideways force (of up to one tonne). The wings transfer this force into the ground.

Wind-powered speed records are unlike normal records, where more power equals more speed. In fact, more wind does not always equal more speed. Instead, a technical solution is required, which maximises lift, but minimises drag (or friction).

The Greenbird is very much like a very high-performance sailboat, but it uses wings instead of sails, and three skates instead of a keel on the hull. It has one vertical wing (similar to a boat's sail, but made of a solid composite instead of canvas), and two horizontal wings (which resemble the wings on a racing car). The Greenbird's vertical wing provides lift exactly like an aircraft wing, except that it pushes the craft forwards, not upwards. Then, to prevent the vehicle from falling over, horizontal wings are used to keep the vehicle close to the ground, much like the wings in a Formula One car.

Once the Greenbird starts to move, it creates an apparent wind, which has tremendous force and can be much faster than the true wind. It acts aerodynamically on the vertical wing and, because the vehicle is light and efficient, there is very little drag.

This technological miracle has produced the fastest wind-powered land vehicle in the world.

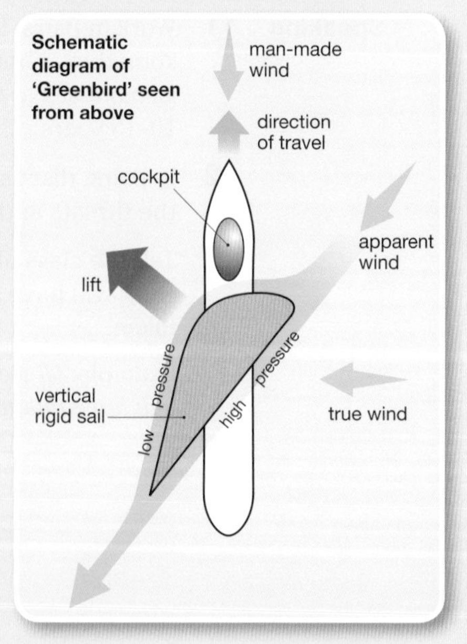

3 Answer these questions about the text.

1 How is the Greenbird *similar to* and *different from* (a) an aircraft, (b) a racing car and (c) a sailing boat? Answer in note form.
2 What is the maximum speed of the Greenbird compared with the wind?
3 What is the maximum force of a side wind that the Greenbird can withstand?
4 What can stop other vehicles going faster even when the wind is stronger?
5 Which of the Greenbird's wings provide a downwards force?
6 What would happen to the vehicle if lift was minimised and drag was maximised?

Vocabulary

4 Notice that the writer of the text in 1 uses the strong adjective *spectacular* (instead of *very good*). Find four strong adjectives like this in the text in 4. Why does the writer use adjectives like these?

Language

Here are some ways of expressing similarity and difference:

Similarity	*The Greenbird relies on a solid sail, which is* ***(very) similar to / (a lot / a little / very much) like*** *an aircraft wing.*
	The horizontal wings ***resemble*** *the wings on a racing car.*
	Just as / In the same way as *airflow over a horizontal wing pushes an aircraft up, the flow of air over the Greenbird's vertical wing propels the vehicle forwards.*
Difference	*Wind-powered speed records are* ***(completely) unlike / different from*** *normal speed records.*
	The Greenbird uses a solid sail ***instead of*** *a canvas one.* ***Instead****, a technical solution is required.*
Similarity + Difference:	*The Greenbird's vertical wing provides lift* ***exactly like*** *an aircraft wing,* ***except that*** *it pushes the craft forwards,* ***not*** *upwards.*

5 Describe these objects in a way which expresses their *similarity to* and *difference from* other objects.

Example: *1 A surfboard is like a small boat, but / except that it uses a flat board instead of a hull.*

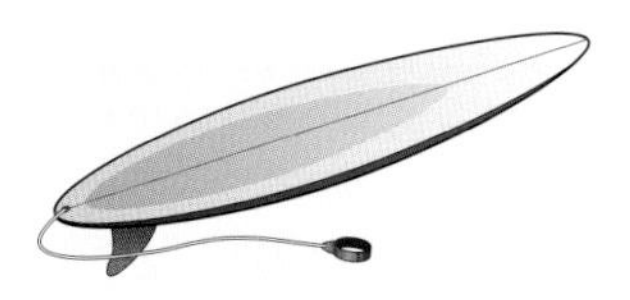

1 A surfboard

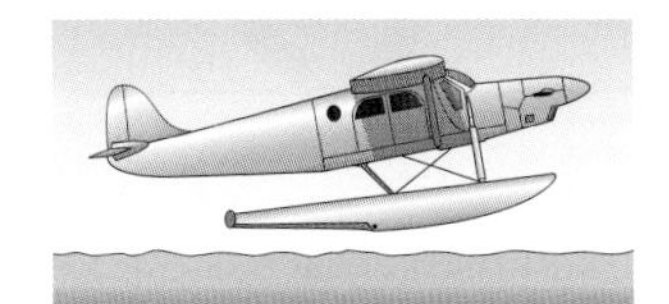

3 A seaplane

5 A submarine

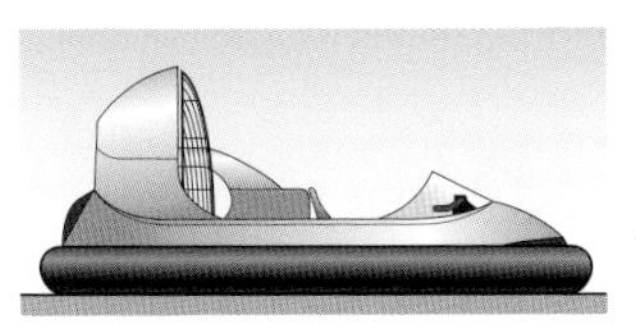

2 A hovercraft

4 A helicopter

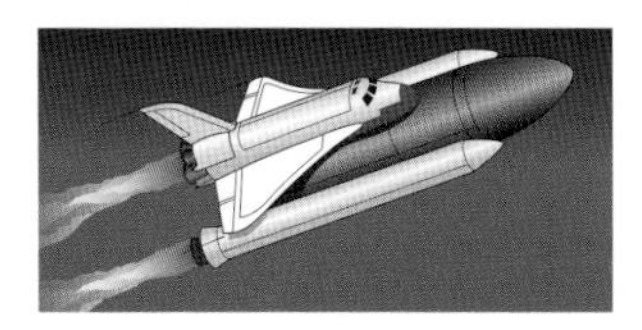

6 A space shuttle

Scanning

6 Practise your speed reading. Look for the information you need on the SPEED SEARCH pages (116–117). Try to be first in your class to do this task.

Task: find out these facts about a new electric car charged by wind power alone.

1 Speed: more than __________ kph
2 Acceleration: faster acceleration than __________ (racing car)
3 Range: capable of driving __________ km before re-charging.

Writing

7 Write a short explanation of how the Greenbird works, using the ideas from your group's discussion in 1. Do not look again at the text in 2.

3 Priorities

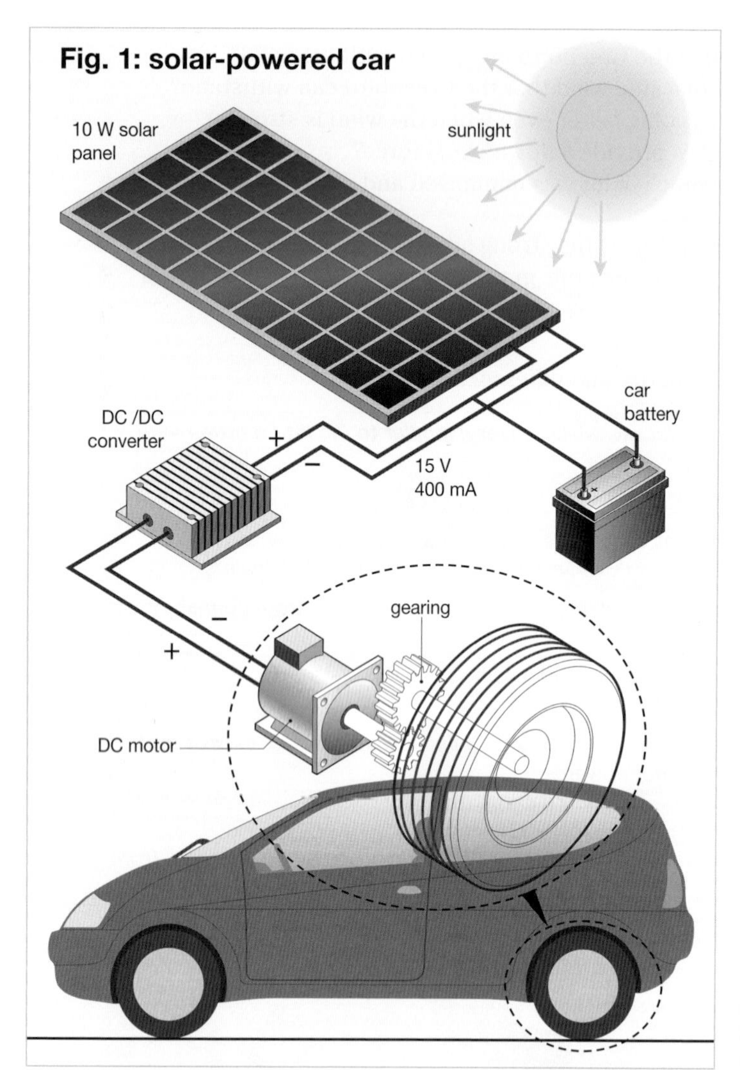

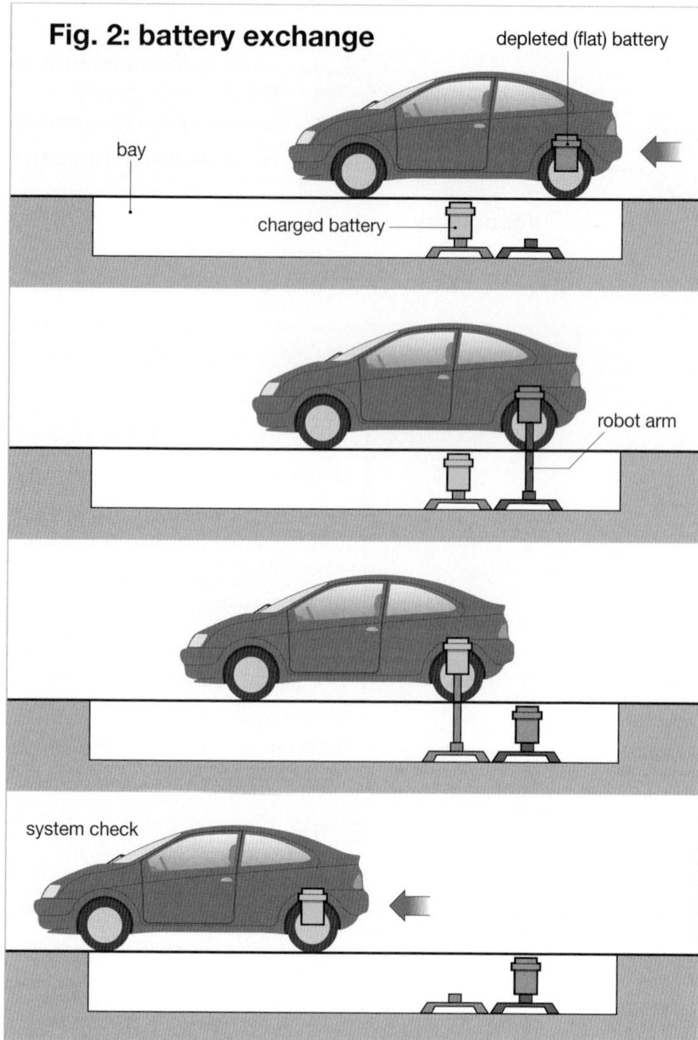

Start here 1 Work in pairs. Discuss the following questions about the four illustrations on this page and the next.

1 In what way(s) are these systems similar?
2 How are they different?

Task 2 Form a group with two other pairs to do this task.

Your group has to decide on the best electric car system for future large-scale use in your country. Prepare for a meeting to decide the order of priority from 1 (best option) to 4 (least good option). Use the information about the four systems on page 113 and follow the instructions below to choose the best option. Then prepare your arguments.

Student A: Argue for the cheapest system to buy.
Student B: Argue for the cheapest system to run.
Student C: Argue for the safest system to use.
Student D: Argue for the simplest and most convenient system.
Student E: Argue for the system that is quickest to refuel/recharge.
Student F: Argue for the system that is least harmful to the environment.

3 Hold the meeting with your group. Make notes of the meeting's agreed order of options, and reasons for the decisions.

Fig. 3: hydrogen fuel cell

A
1
2

B
5
4
3

C
6

Key
electron
hydrogen
oxygen
water

fuel cell: generates electricity
drive unit
fuel tank: stores oxygen

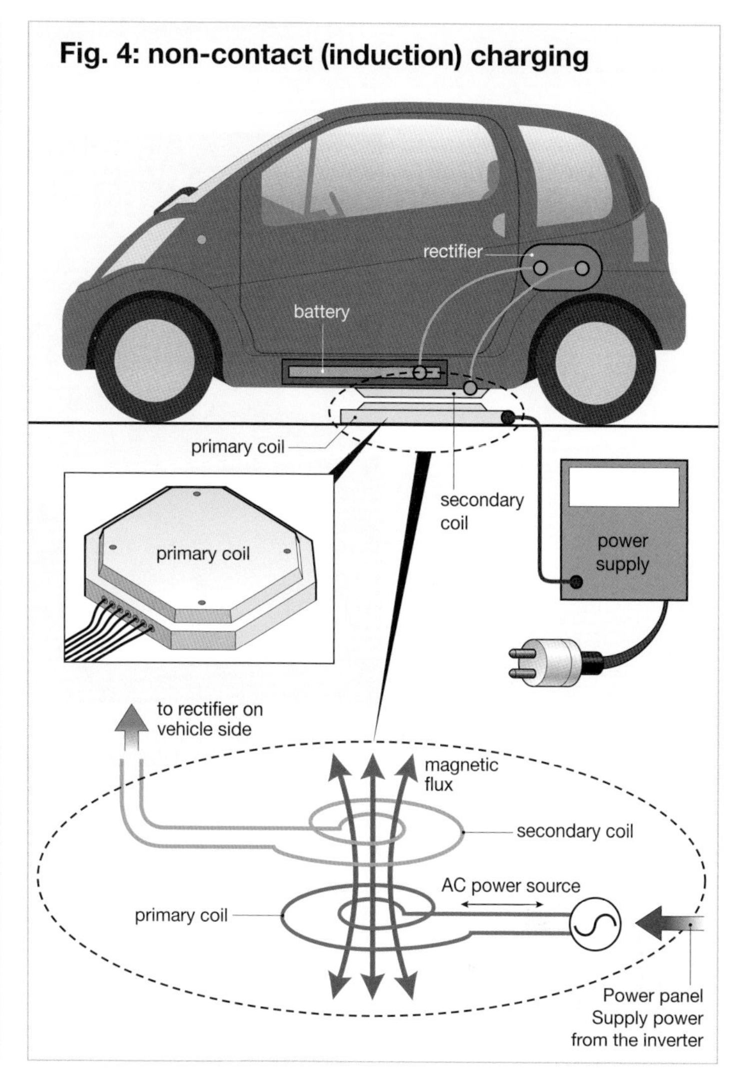

Writing 4 Work individually. Write a report of your meeting, using your notes. In your report, include the following:

1 a short paragraph explaining the background (or reason) for holding the meeting
2 a brief technical description of the four electric car systems (main points)
3 a statement of the decisions you made: from 1 (best) to 4 (least good)
4 a short explanation of the main reasons for your decisions on options

Speaking 5 Prepare for a class (or group) debate: *The best technology for the future environment.* Make a list of technologies (including ones described in various units of this book) which may help to avoid ecological disaster in the future. Put them in order of most helpful to least helpful. Make notes of your reasons.

6 Participate in the debate. Use your notes to help you give reasons for your opinions.

Review Unit F

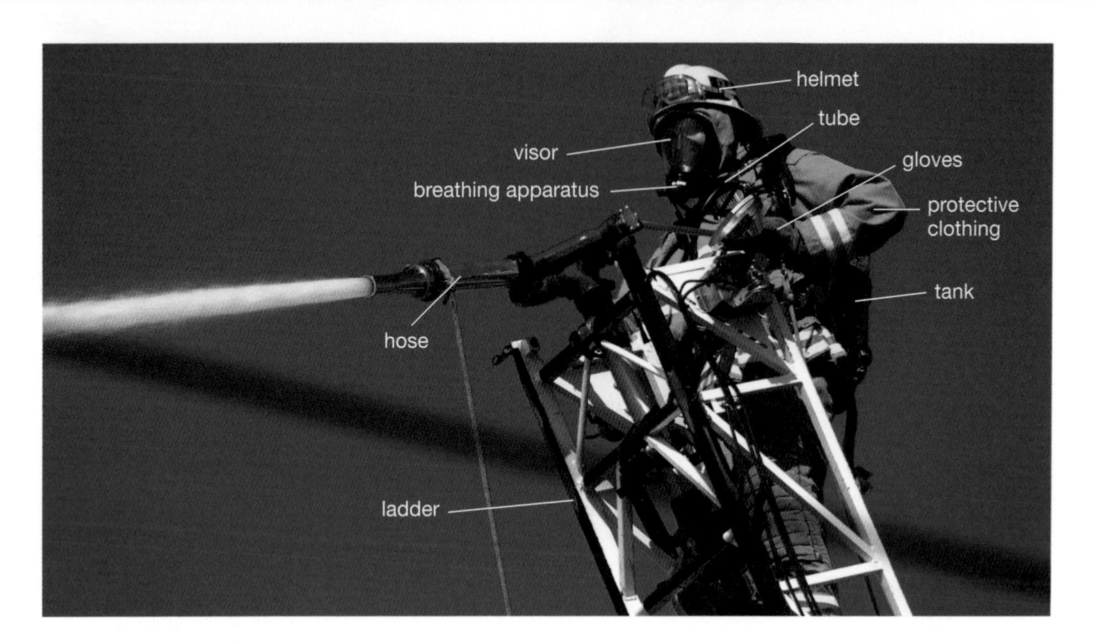

1 Work in pairs. Discuss the firefighter's equipment in the photo. What materials do you think these items may be made of? What are the likely properties of the materials?

2 Rewrite these sentences to give the same meaning, using the words in brackets.

Example: 1 *This plastic does not melt when you heat it.*

1. This plastic is capable of being heated without melting. (melt / when / heat)
2. The frame of this mountain bike is very strong but lightweight. (strength / weigh)
3. Aramid fibre can be pulled with great force without breaking or stretching. (pull / break / stretch)
4. This highly elastic polymer stretches a little when you pull it, and then it can return to its original shape. (stretched / pulled / capable)
5. The soft plastic foam inside this helmet is impact absorbent, and the polycarbonate external shell is impact resistant. (absorbs / able / resist)
6. You can heat the fibre used in this cloth to a high temperature, but it doesn't burn or transfer the heat to the body. (heated / without / burning)

3 Change the word in italics to the word in brackets in these sentences. Make any other changes to give the same meaning.

Example: 1 *The cables in a suspension bridge need to be very strong in tension.*

1. The cables in a suspension bridge need to *have* good tensile strength. (be)
2. The diving suit used by a scuba diver has to *be* totally water resistant. (have)
3. The concrete used in the bridge piers must *be* very strong in compression. (have).
4. The steel used in the axle of a racing car must *have* very good torsional strength and excellent shear strength. (be)
5. We need to design a new running shoe that *is* much more flexible and tough than the old one. (has)
6. Some of the materials used in earthquake-proof building need to *be* slightly elastic. (have)
7. Gold *is* highly malleable and ductile, and *is* highly resistant to corrosion. (has)
8. The fire doors are made of a new material that *has* no flammability and high thermal resistance. (is)

4 Complete these sentences using the words in the box. Use each word once only.

capable	incapable	unable	ability	inability	capacity

1. The report stated that the skyscraper collapsed due to the ____________ of the foundations to support the load.
2. We believe that the server crashed because it was ____________ of handling requests from millions of computers.
3. Fabrikator's newly designed green bus is larger than the old one: it has the ____________ to carry over 70 passengers (instead of 65 in the old one).
4. The fire fighting response unit came too late. As a result, it was ____________ to prevent the forest fire from spreading to the town, and forty houses were destroyed.
5. We'll be able to receive TV when we travel through the desert, because the new communications satellite is ____________ of covering the whole Sahara region.
6. Our company has designed an earthquake-resistant bridge which has the ____________ to withstand earthquakes up to 6.5 on the Richter scale.

5 Study the notes about your client's problems below. Make suggestions to your client to solve them. Use each word or phrase in the box once only.

Why don't you	Let's	Let's try	How about	I would suggest that you	You could

Example: *1 How about trying an open source operating system?*

1. *has problem with computer operating system – solution: use open source*
2. *can't think of ways to improve performance – solution: brainstorm with her*
3. *company has high accident rate – solution: buy new safety equipment*
4. *warehouse loses stock every month – solution: install CCTV system*
5. *needs clarification about my proposals – solution: have meeting*
6. *doesn't know how to reduce energy consumption – solution: convert all factory lighting to low-energy lamps*

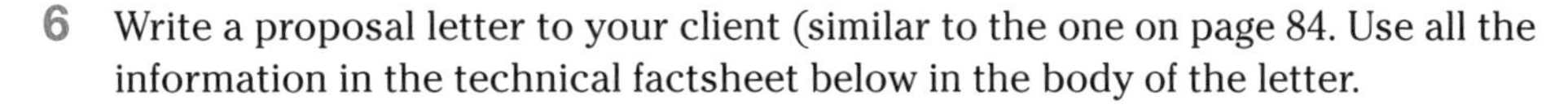

6 Write a proposal letter to your client (similar to the one on page 84. Use all the information in the technical factsheet below in the body of the letter.

You are a technical sales manager with FireProtect, a company which makes and sells fire fighting equipment. After giving a presentation to your client at the fire station, you phoned the client, who invited you to submit a proposal to supply the fire station with the clothing. Invent your client's name, title, company name and address.

PROTECTIVE CLOTHING FOR FIREFIGHTERS
CONTENTS: jacket and trousers
MATERIALS: (1) outer shell – Nomex (2) moisture and thermal barrier (middle layer) – Gore-Tex (3) inner lining: Nomex + viscose
PROPERTIES OF MATERIALS: Nomex – meta-aramid fibre. High heat, flame and chemical resistance. Resists temperatures 350°c +. Material does not melt when exposed to flame. High abrasion resistance, strength, and durability. Gore-Tex – a high-performance weatherproof fabric. Waterproof, windproof, breathable. Viscose – smoothness, softness, comfort.
PRICE: €359.00
FREE DELIVERY: three weeks.

7 Work in pairs. Discuss these photographs and make some predictions for the future. Listen to your partner's predictions, and then agree or disagree with them, giving your reasons.

8 Make predictions for the year 2060 based on the magazine articles below. Use the future perfect.

Example: 1 *By 2060, I think all prisons will have closed down because …*

1

2 July, 2035

27 MORE PRISONS are to close because more biometric tracking and monitoring of offenders are gradually being introduced.

2

13 September **2041**

MALARIA will die out later this year, according to medical researchers. A deadly virus is wiping out all mosquitoes throughout the world.

3

10 MARCH, 2049

A 'NARROW-BAND MODULATED' SIGNAL from a star in outer space was picked up a week ago by a retired electrical engineer with a home radio receiver. The United Nations is now discussing how to make contact with the extra-terrestrial broadcaster.

4

8 June, 2051

A SYSTEM of giant mirrors floating in space is planned within the next five years. Construction of the mirrors will begin next summer. The mirrors will bring daylight to people living in cities in the Arctic Circle.

5

23 August, 2055

THE COMPUTER CHIP industry is dying out, says a think-tank report today. Transistors have shrunk in size to the size of molecules, but they can't get any smaller, because silicon transistors are unstable at the molecular level. Computer scientists are now investigating new technologies, such as optical computers (using laser beams).

9 Compare these pairs of photographs, pointing out similarities and differences between the new and old designs.

Example: 1 *The new lathe turns curved shapes just as the old one does, but instead of …*

1 The DMG lathe: main features

- complete safety shield with large safety windows
- computer-controlled
- control panels and 19-inch monitors – can be rotated and swivelled
- password needed for access
- optional seat for operator

2 Crown RC 5500 fork lift truck: main features

- comfortable for driver
- huge operator compartment
- excellent visibility
- top of compartment acts as desktop
- document clip and storage area for tools

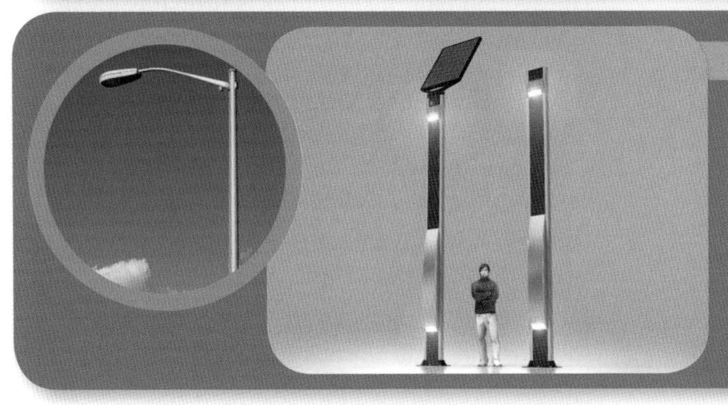

3 The ZIPlux outdoor lamp post: main features

- light source located at base of post
- light transmitted to top by fibre optics
- angle and spread of light can be controlled
- one light source can illuminate different spots
- maintenance carried out at ground level – no ladders needed
- can use solar energy

10 Describe a device or piece of equipment you know about, which has recently been updated, modified or improved in some way.

- Draw a simple sketch of the device/equipment. If possible, draw a plan view and an elevation view.
- Explain its purpose or function.
- Describe its shape and appearance, referring to your drawing(s).
- Explain how it works.
- Explain how it differs from earlier models or designs.
- Make some predictions about how it will have been improved by some future date.

Projects 11 Do the following tasks:

1 Research the future of your industry. Write it up as if you are a 'futurologist'. See page 11 for reference.

2 Make predictions about the properties of new materials that may be developed in the future. Explain the purposes for which they may be needed.

Language summary

A Grammar

Present simple
The present simple is used to talk about:
- regular or routine events: *Hans works with robots every day.*
- job descriptions: *The chief electrician supervises a team of four electricians.*
- processes: *The water flows from the tank into the solar water panel.*

Present continuous
The present continuous is used to talk about:
- activities while the speaker is speaking: *Look. I'm taking the wheel off now.*
- current activities: *I'm studying electronics this year.*
- planned future activities: *I'm doing a Master's degree next year.*

Future
will and *won't* are used to talk about things that you think are certain to happen (or not happen) in the future: *I won't see you next week because I'll be on a training course. We will use more electric cars in 2025.*

The present continuous is used to talk about future events which are already planned: *The whole team is travelling to the conference tomorrow morning.*

going to + verb is used to talk about your (or other people's) intentions: *We're going to present the new prototype at the meeting.*

The present continuous and *going to* + verb are often used interchangeably without any distinction between intentions and plans.
to is used after verbs such as *plan, intend, want, hope*: *I want to deliver the blueprints before the deadline. Hans intends to study at Munich next year. We hope to complete the investigation soon.*

Past participle
The past participle (e.g. *opened, burnt, flown*) is the verb form used by a number of verb constructions, such as the passive and the present perfect.

In regular verbs (which use *-ed*), the past participle is the same form as the past simple: *ejected, pushed.*

In irregular verbs, sometimes the past participle is the same form as the past simple, and sometimes it is a different form: *bend, bent, bent; break, broke, broken.*

A list of irregular verbs where the form changes can be found on page 104.

Present simple passive
The present simple passive uses *is/are* + past participle.

In an active sentence, the subject is the same as the agent. The subject does the action:

A rotating screw (subject = agent) *pushes the plastic pellets* (object).

However, in a passive sentence, the subject is NOT the same as the agent. The subject doesn't do the action. The agent does the action to the subject:

The plastic pellets (subject) *are pushed by a rotating screw* (agent).

The passive is often used in technical writing for two main reasons.

1 The passive can make the writing clearer:

The mould cools the plastic. Then a mechanism ejects the plastic. (active)
➔ *The plastic is cooled by the mould. Then it is ejected.* (passive)

The second pair of sentences is probably clearer than the first because the subject (*plastic/it*) refers to the same topic in both sentences. The passive allows the writer to have the same topic as the subject of two or more sentences in a paragraph. This helps to unify all the sentences of the paragraph around a single topic.

2 The passive helps the reader to focus on actions rather than agents:

Then a mechanism ejects the plastic. (active) ➔
Then the plastic is ejected. (passive)

The second sentence focuses the reader's attention on the action of *ejecting* and indicates that it is not important what (or who) carries out the action. Using the passive allows the writer to omit the agent completely from a sentence. This focuses attention on the topic (*plastic*) and the action (*ejection*).

Present perfect
The present perfect uses *have/has* + past participle: *We've finished the project. Have they repaired the car? He hasn't found the fault yet.*

The present perfect is used to talk about events during a period of time lasting from the past right up to the present time. You can use *for, since, just, already* and *yet.*

- *for* focuses on the length of an unfinished period of time: *Hans has worked as an engineer for two years.* (Hans is still working as an engineer.)
- *since* focuses on the starting point of an unfinished period of time: *Hans has worked as an engineer since 2012.* (Hans is still working as an engineer.)
- *just* emphasises that the event happened recently: *The aircraft has just landed.* (This happened perhaps one or two minutes before now.)
- *already* (in positives) emphasises the completion of an action before the present: *The builders have already installed the wiring.*
- *yet* (in questions and negatives) emphasises the period of time to the present: *Have you changed the brake fluid yet? He hasn't changed the brake fluid yet.* (After all this time, the brake fluid is still unchanged.)

Past simple
The past simple is used to talk about events which happened at a specific time in the past. *I went yesterday. They arrived this morning. When did you leave? He didn't complete the report by the deadline.*

You can use time expressions like these with the past simple: *yesterday, this morning, the day before yesterday, three minutes ago, two days ago, five weeks ago, last week, last month, last year, in 2009, on 20th October, at 6.30.*

Present perfect v past simple
Here is an example of the difference in use between the present perfect and past simple:

- present perfect (no specific time is mentioned or implied): *The Ares rocket has crashed.*
- past simple (a specific time is mentioned or implied): *The Ares rocket crashed early this morning.*

Here is another example of the difference between the present perfect and the past simple. This is part of Reme's CV. Imagine that the present day is in the year 2015.

Apprentice at MultiPlastics: 2008–2011 Junior technician at MultiPlastics: 2011–present day

Past simple: *Reme was an apprentice at MultiPlastics for three years. She worked as an apprentice from 2008 to/until 2011. She joined MultiPlastics seven years ago.*

Present perfect: *Reme has worked at MultiPlastics for seven years. She's had a job as a junior technician since 2011. She's been at MultiPlastics from 2008 until now.*

Present perfect passive
The present perfect uses *have/has* + *been* + past participle: *The project has been completed. Has the car been repaired? The fault hasn't been found yet.*

The reasons for using the passive with the present perfect are the same as those for using the passive with the present simple (see page 100).

The present perfect (active and passive) is often used with *just, already* and *yet.*

The spar has just been constructed.
Five risers have already been built / have been built already.
Has the topside been fitted to the spar yet? No, the work hasn't been done yet.

Past simple passive
The past simple passive uses *was/were* + past participle. *The project was completed last year. The car was repaired this morning. The fault wasn't found when the car was serviced.*

The reasons for using the passive with the past simple are the same as those for using the passive with the present simple (see page 100).

In the examples above, the passive focuses on the *action (verb)*. The *agent* is omitted to show that we don't know (or are not interested in) who (has) carried out the action.

Question forms: *Has the topside been fitted? When was it fitted? Have the risers been built? When were they built?*

Present perfect passive v past simple passive
Here is an example of the difference in use between the present perfect and past simple.

- present perfect passive (no specific time is mentioned or implied): *The topside has been fitted to the spar.*
- past simple passive (a specific time is mentioned or implied): *The topside was fitted to the spar in 2007.*

Past continuous
The past continuous uses *was/were* + the *-ing* form of the verb. *The passenger was walking through the security checkpoint.*

The past continuous is used to talk about situations in progress in the past. It is often combined with a past simple for a shorter action, using *while, when* or *as.*

In the next two examples, the situation in progress is underlined, and the ***short action*** is in bold.
While/When/As his case was going through the X-ray machine, ***he walked through the WTMD.***
Or: ***He walked through the WTMD*** *while/when/as his case was going through the X-ray machine.*

In the next two examples, the short action was sudden, or interrupted the situation in progress. You can use *while*, *when* or *as* with the situation in progress. You can only use *when* with the short action.

While/When/As his case was going through the X-ray machine, ***the alarm sounded.***

His case was going through the X-ray machine ***when the alarm sounded***.

In the next example, two situations were in progress simultaneously over the same period of time. It is possible to use *when* here, but you usually use *while* or *as* to emphasise the element of progression.

- While/As the cases were going through the X-ray machine, the officials were studying the screen.

Future perfect

The future perfect active uses *will/won't* + *have* + past participle. The future perfect passive uses *will/won't* + *have* + *been* + past participle.

The future perfect is used to make a prediction about an action or situation that will be finished before (or *by*) a specified time in the future.

- active: *By 2060, the sea level will have risen 1.2 metres.*
- passive: *By 2060, many low-lying countries will have been flooded.*

In these examples, the rise in sea level and the flooding happen before 2060.

Modal verbs v semi-modal verbs

- *Must* and *should* are modal verbs. They don't take -*s* in the 3rd person. They don't use *do/does* in the question or negative: *Everyone must wear protective clothing on this site. Should she put on her safety goggles now? He mustn't take off his safety boots until he leaves the site.*
- *Have to* and *need to* are semi-modal verbs. They are like modals in meaning, but have the same form as normal verbs. They take -*s* in the 3rd person. They use *do/does* in the question or negative: *Everyone has to wear protective clothing on this site. Does she need to put on her safety goggles now?*

The negative forms of *have to* and *need to* have a different meaning from the negatives of *must* and *should*. For example, *she doesn't have to take off her safety goggles* means *it is unnecessary for her to take off her safety goggles.*

- *You must / need to / have to do it* = it is essential (necessary) to do it.
- *You mustn't do it* = it is essential (necessary) not to do it.
- *You should do it* = it is recommended that you do it.
- *You don't have to / don't need to do it* = it is unnecessary to do it.

Modal verb + passive infinitive

Must, should, have to and *need to* are often followed by passive infinitive verbs (*be* + past participle) in safety rules and procedures: *Hard hats must be worn at all times. This package has to be kept cold. Mobile phones should not be used here.*

Future modals

Verbs like *can, must, have to* and *need to* normally refer to both present and future time. However, to emphasise that they refer to the future, you can use *will be able to, will have to* and *will need to.*

- *can* ➔ *We'll be able to fly to Mars by 2050, but we won't be able to fly to Jupiter by then.*
- *must / have to / need to* ➔ *The engineers will have to / will need to complete the bridge by the end of next year, but they won't have to / need to complete the road by then.*

Be going to can be used instead of *will*: *We're not going to be able to do it by 2019. You're not going to have to do it by 2019.*

There is no future form of *mustn't*, so to express the future other expressions with similar meanings should be used: *He mustn't do it.* ➔ *He'll have to make sure he doesn't do it.*

We can use *be going to* instead of *will*. Examples: *We're not going to be able to do it by 2019. You're not going to have to do it by 2019.*

Conditionals

- The first conditional uses *if* + present simple + *will/won't*: *If the aircraft engine fails, the pilot will activate the ejection system.* The first conditional is used to talk about things which are likely to happen in the future under certain real or possible conditions.
- The second conditional uses *if* + past simple + *would/wouldn't*: *If my car had a hydrogen fuel cell, it wouldn't emit any carbon into the atmosphere.* The second conditional is used to talk about things which are unlikely or impossible to happen in the future, because the condition is unreal, or imagined. In the above example, the car doesn't have a hydrogen fuel cell, so the condition is unreal. Notice that the past simple in the condition refers to the present time.
- The third conditional uses *if* + past perfect* + *would/wouldn't* + perfect infinitive: *If he had repaired the brakes, his car would have stopped in time. If he had not forgotten to check his brakes, the accident would not have happened. If the brakes had been repaired, the accident would have been avoided. If the brakes had not been worn, the driver would have been able to stop in time.* The third conditional is used to speculate about possible past events that did not in fact happen. In the

above examples, the car's brakes were worn and were not repaired, and the accident actually happened. The speaker is imagining an alternative past. The present infinitive can be used instead of the perfect infinitive to emphasise an imagined present result: *If he had repaired the brakes, his car would be fine now / he would be driving around in it now.*

Note: in all types of conditionals, the *if* clause can come first (as in the examples above) and then a comma is usually used after the clause. However, it can also be the second clause: *My car wouldn't emit any carbon into the atmosphere if it had a hydrogen fuel cell.*

* The past perfect uses *had* + past participle.

Reported statements

Reported statements use a reporting verb (such as *say, tell, explain*) which can be followed (optionally) by *that*: *He said (that) he had fixed the car.*

After a past tense reporting verb (such as *said, told, explained*), the verb used in the actual words usually shifts one tense into the past: present ➔ past; present perfect ➔ past perfect; *can* ➔ *could*; *will* ➔ *would*, and so on.

Pronouns and possessive adjectives also change, e.g. *I, you* ➔ *he/she*; *we* ➔ *they*; *me* ➔ *him/her*; *your* ➔ *his/her*. Some other words also change, e.g. *come* ➔ *go*; *this* ➔ *that*; *yesterday* ➔ *the day before*; *tomorrow* ➔ *the next day*.

He told me (that) he was a policeman. ("I am a policeman.") She told him (that) he could go. ("You can go.") He told him (that) he hadn't searched his bag. ("I haven't searched your bag.") She explained (that) she had been an apprentice the previous year. ("I was an apprentice last year.") He said (that) he would go there again the next day. ("I'll come here again tomorrow.")

The tense shift is not always necessary if the context is clear. For example (in the text in 2 on page 52): *He explained that the surgeon had put a metal plate inside his jaw, which made the metal detector beep.* Here the verb *made* does not shift back.

Reported instructions

Reported instructions use an instructing verb (such as *order, tell, instruct*) followed by *(not) to*.

The official told him to open his bag. ("Open your bag.") The policeman instructed him not to pass through. ("Don't pass through.")

Modifying comparative adjectives

Comparative adjectives can be modified in a general way using *far, much, a great deal, a lot, slightly, a little*: *This motorbike is far more / much more / a great deal more / a lot more / less expensive than that one. This new model of car is only slightly better / a little better than the old one.*

A comparison can be modified in a more specific way like this: *The new hard disk is three times / ten percent / a third / 3 MB larger than the old one. This engine is twice / three times / half as powerful as that one.*

Note: *five times larger than* = *five times as large as*. But you can't say ~~*twice larger than*~~ or ~~*half larger than*~~.

Comparative adverbs can be modified in a similar way: *This car accelerates far / much / a great deal / a lot / slightly / a little less quickly / more slowly than that one.*

Modifying superlative adjectives

Superlative adjectives can be modified using *easily* and *by far*. *Air travel is easily / by far the fastest form of public transport.*

Non-defining relative clause

A non-defining relative clause is a useful way to join two sentences together. It does not provide part of a definition, or limit the meaning of the preceding word. It simply adds further information.

The non-defining relative clause uses relative pronouns such as *which, who, where* and *from where*. A comma is used immediately before the relative pronoun.

Non-defining relative clauses are most commonly used in written English.

Plastic pellets are stored in the hopper, which then feeds them slowly into the extruder. The goods were shipped to the warehouse, where they were kept securely for some weeks. This is the main transmitter, from where all the signals are sent. This office belongs to the chief technician, who is in charge of all the servers.

No comma is used immediately before the relative pronoun.

Defining relative clause

A defining relative clause limits the meanings of the preceding words and is often used in definitions: *A transmitter is a device which sends signals.*

Word Parts

Note: Sometimes the word parts have different meanings. Check new words in a dictionary.

Word part	Usual meaning	Example
aero-	air	aeronautical
aud-	hearing	audible
centi-	hundred(th)	centilitre
ex-	out / out of	extrusion
-en	causation	strengthen
frig-	cold	refrigerant
geo-	Earth	geostationary
hemi-	half	hemisphere
hydr-	water	hydrostatic
ign-	fire	ignition
in-	into	injection
in-	not	incapable
inter-	between	intercom
intra-	inside	intranet
kilo-	thousand	kilometre
-less	without	wireless
lubr-	oil	lubricant
max-	large	maximise
-meter	measuring	voltmeter
micro-	very small	micro-manage
mini-	small	minimise
multi-	many	multi-storey
non-	not	non-contact
poly-	many	polycarbonate
-proof	preventing	flame-proof
re-	again, back	rebuild
semi-	half	semicircular
sol-	sun	solar
sub-	under	sub-sea
super-	much greater	supercapacitor
tele-	distant	telecommunications
therm-	heat	thermoplastic
trans-	across	transmission
tri-	three	triangular
ultra-	very much greater	ultracapacitor
un-	not	untie
vis-	seeing	visible

Irregular verbs

Here is a brief list of irregular verbs whose past simple form is different from their past participles.

become	became	become
break	broke	broken
do	did	done
drive	drove	driven
fall	fell	fallen
fly	flew	flown
go	went	gone
rise	rose	risen
run	ran	run
speak	spoke	spoken
take	took	taken
tear	tore	torn
wear	wore	worn
write	wrote	written

B Functions and notions

Method

You can talk about *method* (= *how to do something*) in these ways.

- *by* + *-ing* form:

The passenger activates the machine by touching the screen. The spar was secured by fastening it to the seabed with nine cables.

- *(by) using / by means of* + noun:

The car avoids collisions automatically (by) using a camera in its bumper. The spar was secured to the seabed by means of nine cables.

Questions about method: *How do you do it? How did you do it? How is it done? How was it done?*

Aim, purpose or objective

You can talk about the purpose of an action in these ways.

- You can use (*in order*) *to* + verb to talk or write about the purpose of an action: *The spar was secured (in order) to prevent it from moving around in heavy seas. Why do you put steel bars in the concrete? To reinforce it. Steel bars are put in the concrete to reinforce it.*
- *The aim/purpose/objective of painting the car body is to protect it from rust.*
- Combined with *if/whether*: *The aim/purpose/objective of the investigation was to find out whether the collapse of the bridge was due to corrosion.*

Questions about aim/purpose/objective: *Why do/did you do it? Why is/was this done? What is/was the objective/aim of doing it?*

Appearance
You can describe the shape or appearance of something in various ways: *It looks like a gherkin. It's T-shaped. It's in the shape of a star. It's an L-shaped building. It's circular/oval/square (in shape).* Questions about appearance: *What shape is it? What does it look like?*

Degrees of certainty
You can express certainty like this: *It's certain that they will make a plastic fuselage one day. They will certainly/ definitely not make a plastic engine before 2035.*

You can express probability like this: *It's probable/ likely that they will make a plastic fuselage before 2035. They will probably not make a plastic fuselage next year. It's unlikely that they will do it. (= It's likely that they won't do it.)*

You can express possibility like this: *It's possible that they will make a plastic wing one day. They will possibly make a plastic wing one day.*

Speculating about the past
You can use these language forms to speculate about the past where you don't have all the information.

- *may/might/could* + perfect infinitive to indicate possibility: *One of the bridge cables may/might/ could have broken.* (= *I think it's possible that one of the bridge cables broke / has broken.*)
- *must* + perfect infinitive to indicate certainty: *One of the bridge bearings must have corroded.* (= *I'm certain that one of the bearings (has) corroded.*)
- *can't / couldn't* + perfect infinitive to indicate impossibility: *The storm can't / couldn't have caused all this damage.* (= *I think it's impossible that the storm caused all this damage.*)

The passive uses *may/might/could/must/can't / couldn't* + *have* + *been* + past participle: *All this damage can't have been caused by the storm.*

Criticising
Should + perfect infinitive can be used to criticise a past action that was wrongly taken, or not taken: *Your company should/shouldn't have replaced the bearings.* (active) *The bearings should/shouldn't have been replaced.* (passive)

Making suggestions
You can make suggestions using a variety of forms: *Try installing a new anti-virus program. Why don't you build a fence around the warehouse? You could replace the old computers with some new ones. I (would) suggest that you hire a new security guard.*

When your suggestion is for yourself and others in a group, you can use these forms: *Let's work on this project together. Let's try brainstorming for five minutes. Why don't we have a meeting about this?*

Questions asking for suggestions: *What should I/we do? What do you suggest I do? What do you think we should do?*

Describing properties
You can describe the properties of materials in a variety of ways.

- present simple active: *This material resists heat. The material bends easily.*
- *can/can't*: *You can stretch it. It can be stretched. The material can be bent easily.*
- *without* + *-ing* form: *The material can absorb a blow without bending* (= *… but it doesn't bend*).
- *resist/resistant/resistance*: *This material resists water very well. It is highly water resistant. It is an extremely water-resistant material. It has good resistance to water. It has excellent water resistance.*
- *able/ability/unable*: *This material is able/unable to resist heat. It has the ability to resist heat.*
- *capable/capability/capacity/incapable*: *This plastic is capable/incapable of resisting chemicals. This plastic has the capability to resist / of resisting chemicals. This steel has the capacity to resist bending.*
- suffixes *-able/-ible/-proof*: *flammable, fusible, waterproof.*

A hyphen (-) is normally added when an adjectival phrase goes in front of a noun and becomes a single hyphenated adjective.

Noun phrases such as *resistance to water* and *water resistance* mean the same and can be used interchangeably.

Cause and effect
You can express the cause of something in a number of ways:

- *due to / owing to / because of / caused by / as a result of* + noun phrase: *The pipes burst because of / owing to / due to / as a result of the high water pressure. The Earth's temperature is rising owing to / as a result of the increase in carbon emissions.*
- verb suffixes: *-ify*, *-efy*, *-ise (-ize)*, *-en*: *gasify* (= *turn something into a gas*), *liquefy* (= *turn something into a liquid*), *equalise* (= *make things equal*), *strengthen* (= *make something stronger*).

Sequence
You can talk about the sequence of events in a number of ways: *The capsule reaches a safe altitude, (and) then the parachutes open up. After / Once / When / As soon as the capsule reaches / has reached a safe altitude, the parachutes open up.*

The order can be reversed: *The parachutes open up after the capsule has reached a safe altitude.*

As soon as emphasises that the second action happens very quickly after the first one.

If both clauses have the same subject, *after* can be followed by the present participle (verb ending in -*ing*): *After detecting a problem, the computer sends a warning.* (= *After it has detected a problem, the computer sends …*)

Comparison
Two items can be compared in these ways.

- comparative adjective + *than*: *The modified barrel is narrower than the standard one.*
- *more/less* + adjective/noun: *It's also more accurate and has less recoil.*
- *but/while/whereas*: *It's straight, but/while/whereas the other one is conical.*

Instructions: *only … if* v *don't … unless*
Some rules and instructions can be expressed in both a positive and a negative form. Positive: *Only change your heading if/when/after you've checked your flight level.* Negative: *Don't change your heading unless/ until you've checked your flight level. Don't change your heading without/before checking your flight level.*

Deadlines
A *deadline* (the last possible date for doing a task) can be set in these ways: *Please finish the report by the end of June (at the latest). We have to start the project no later than 15th March. The deadline for completing the investigation is the end of the year.*

Expressing similarity and difference
Here are some ways of expressing similarity: *A sail is (very) similar to / (a lot / a little / very much) like an aircraft wing. The horizontal wings resemble the wings on a racing car. Just as / In the same way as airflow pushes a horizontal wing upwards, (so) airflow pushes a vertical sail forwards.*

Here are some ways of expressing difference: *Wind-powered speed records are (completely) unlike / different from normal speed records. The Greenbird uses a solid sail instead of a canvas one.*

We can combine similarities and differences, as in the following examples:

The Greenbird is very much like a sailboat, but / except that it uses wings instead of sails.

The vertical wing is like a boat's sail, but made of a composite instead of canvas.

Reference section

1 Abbreviations

Prefixes to SI units

n-	**nano-**	(10^{-9})	(e.g. nanometre: nm)
µ-	**micro-**	(10^{-6})	(e.g. micrometre: µm)
m-	**milli-**	(10^{-3})	(e.g. millimetre: mm)
c-	**centi-**	(10^{-2})	(e.g. centimetre: cm)
d-	**deci-**	(10^{-1})	(e.g. decilitre: dl)
k-	**kilo-**	(10^{3})	(e.g. kilometre: km)
M-	**mega-**	(10^{6})	(e.g. megawatt: MW)
G-	**giga-**	(10^{9})	(e.g. gigabyte: GB)
T-	**tera-**	(10^{12})	(e.g. terabyte: TB)

Note: µ (pronounced 'miu') is the Greek letter *m*.

Length

mm millimetre(s)
cm centimetre(s)
m metre(s)
km kilometre(s)

Area

mm² square millimetre(s)
m² square metre(s)
km² square kilometre(s)

Volume/Capacity

mm³ cubic millimetre(s)
cm³ cubic centimetre(s)
m³ cubic metre(s)
km³ cubic kilometre(s)
ml millilitre(s)
cl centilitre(s)
l (or L) litre(s)

Note: In AmE, the spellings *(-)meter(s)* and *(-)liter(s)* are used: *meter, kilometer, liter, milliliter.*

Mass/Weight

mg milligram(s)
g gram(s)
kg kilogram(s)
t tonne(s)
kt kilotonne(s)
Gt gigatonne(s)

Time

s second (also sec.)
min minute (also m, as in rpm)
h hour

Electricity

A ampere(s) or amp(s)
MA mega amp(s)
Ah ampere hour(s)
W watt(s)
kW kilowatt(s)
kWh kilowatt hour(s)
MW megawatt(s)
V volt(s)

Frequency

Hz hertz
kHz kilohertz
MHz megahertz

Pressure

Pa pascal(s)
kPa kilopascal(s)
bar bar (= 100 kPa, atmospheric pressure on Earth at sea level)

Force

N newton
Nm newton metre (a measurement of torque)
G 1G = the force of the Earth's gravity

Sound level

dB decibel(s)

Speed/Rate

m/s metre(s) per second
km/s kilometre(s) per second
km/h kilometre(s) per hour
kph kilometre(s) per hour
rpm revolution(s) per minute

Digital information storage

kB kilobyte
MB megabyte
TB terabyte

Other abbreviations and units in common use

ft foot/feet (= 0.3048 m)
in inch(es)
gal gallon(s)
pt pint(s)
yd yard(s)
mi mile(s) (m is also used)
mph mile(s) per hour
mpg mile(s) per gallon
gph gallon(s) per hour
psi pound(s) per square inch
g/km gallon(s) per kilometre
lb pound(s)
oz ounce(s)
cc cubic centimetre(s) (engine capacity)
cu m cubic metre(s)
sq m square metre(s)
ppm parts per million
yr year
Gt C/yr gigatonne(s) of carbon per year
L/kg litre(s) per kilogram

Temperature

°C degree(s) Celsius
°F degree(s) Fahrenheit

Some other abbreviations used in this book

ABS	anti-lock braking system
AC	alternating current
ACC	adaptive (or autonomous) cruise control
approx.	approximately
b/d	barrels (of oil) per day
BEng	Bachelor of Engineering (degree)
cc	(document) copied to; cubic centimetres (engine capacity)
CCS	carbon capture and storage
CCT	clean coal technology
CCTV	closed-circuit TV (system)
CFL	compact fluorescent light (bulb)
CH_4	methane
CO	carbon monoxide
CO_2	carbon dioxide
CPR	cardio-pulmonary resuscitation (emergency procedure for someone who has stopped breathing and has no pulse)
CV	curriculum vitae, a summary of skills, qualifications and work experience
CWS	collision warning system
d	depth (e.g. *d: 20 mm*)
DC	direct current
DNA	deoxyribonucleic acid
DVD	digital video disc
E	east (90°)
e.g.	for example
enc.	enclosed (or attached) document
EPP	expanded polypropylene
ESC	electronic stability control
etc.	and so on / etcetera
EV	electric vehicle
FAQ	frequently asked questions
Fig.	figure
FL	flight level (1FL = 100ft)
FYI	for your information
GEOSAR	geo-stationary search and rescue (satelllite)
GPS	global positioning system
HQ	headquarters
HHMD	hand-held metal detector
HRU	hydrostatic release unit
h	height (e.g. *h: 100 mm*)
i.e.	that is; in other words
IPAS	intelligent parking assist system
IT	information technology
LAS	launch abort system
LED	light-emitting diode
LEOSAR	low earth orbit search and rescue (satellite)
LKA	lane keeping assist (system)
MP	megapixels (one million pixels); also **Mp**
N, NW, NE	north (0°/ 360°), north west (315°), north east (45°)
N_2O	nitrous oxide
n/a	not applicable
no.	number
O_3	ozone
PV	photovoltaic
ref.	reference / with reference to
S, SW, SE	south (180°), south west (225°), south east (135°)
SatNav	satellite navigation (system)
SO_2	sulphur
SWOT	strengths, weaknesses, opportunities, threats
TPU	thermoplastic polyurethane
v	versus; compared with (also **vs**)
w	width (e.g. *w: 30 mm*)
W	west (270°)
WTMD	walk-through metal detector

2 British and American English

Here are some of the words used in this book, but there are many more. You can find more at the back of *Longman Technical English Levels 1 and 2.* Key the words *American British English* into an internet search engine or *Wikipedia* to find more examples.

Differences in spelling

British English	American English
aluminium	aluminum
calliper	caliper
doughnut (*shape*)	donut
fibre, fibreglass	fiber, fiberglass
gauge	gage
litre	liter
metre	meter
mould	mold
stabilise*	stabilize*
storey (*building*); pl: storeys	story; pl: stories

* *-ise / -iser / -isation* is normally BrE and *-ize / -izer / -ization* is normally AmE but also BrE.

Differences in words

British English	American English
aeroplane*	airplane*
crude oil	petroleum
disc (*brake*)	rotor
earth (*electricity: earthed*)	ground; grounded
ejector seat	ejection seat
flat (*battery*)	dead
indicators (*cars*)	turn signals, blinkers
lift (*building*)	elevator
mobile phone	cell phone
motorway (*roads*)	freeway
petrol (*refined oil*)	gasoline, gas
petrol station	filling station
power station	power plant
socket (*elec*)	jack, outlet

**aircraft* is commonly used in both AmE and BrE

Extra material

3 Events 3 Sequence (2)

Task exercise 5 page 25

Student B

The second part of the ejection sequence (0.5–4 seconds)

1 The rocket lifts the seat up to ________ metres and then ________.	0.5 sec
2 Then a drogue gun, activated by a ________, fires a metal ________.	0.52 sec
3 A small parachute is pulled out of the ________ by the ________.	
4 Once it is out of the seat, the small parachute ________.	
5 At 14500–11500 feet, the ________ pulls out the main ________.	2–4 sec
6 After this, the seat and leg ________ are released.	
7 The main parachute then opens and ________________.	
8 You then ________________ safely.	

9 Design 1 Inventions

Speaking exercise 5 page 69

Student A

Main design features of LED bulb (compared with standard design)

- *Same shape and form as incandescent bulb*
- *Screws into lamp socket like normal lamp*
- *LED light source (not fluorescent or incandescent)*
- *Safer than fluorescent or incandescent bulbs because it has solid-state components*
- *Lasts 10 x longer than fluorescent bulb*

8 Projects 3 Drilling

Speaking exercise 9 page 63

Student A

FACT SHEET STAGE 1A:
Drilling the pilot hole

Purpose: to drill a pilot hole with a small diameter
Drilling procedure

1. attach narrow-diameter drill bit to drill string
2. drill narrow pilot hole
3. enlarge pilot hole using wash pipe
4. pump drilling fluid through drill string to drill bit
5. drill bit + drilling fluid (under high pressure) break rock into small pieces
6. drilling fluid carries rock pieces back to entrance hole at drill rig
7. drill bit reaches exit point and comes out of ground
8. detach drill bit from string and attach reamer

11 Materials 3 Properties (2)

Task exercise 8 page 89

Student A

You want to spend reduced budget on:
Flite and Marathonite running shoes

Reasons:

- *sprinting team is excellent; could get five or six golds*
- *marathon runner has good chance of gold*
- *rain will be a problem in marathon: need slip-resistant shoes*
- *runners have more chance of golds than swimming, sailing and rowing teams*
- *sensors and Doppler lidars expensive*
- *Olympic committee may ban use of SpeedShark swimsuits, so swimsuits may be waste of money*

5 Safety 3 Rules

Speaking exercise 6 page 41

Group A

Rules of the Air (Part 2): Overtaking or passing another aircraft

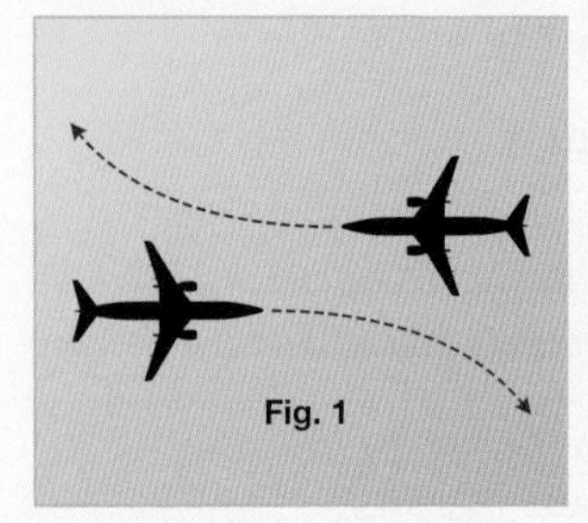
Fig. 1

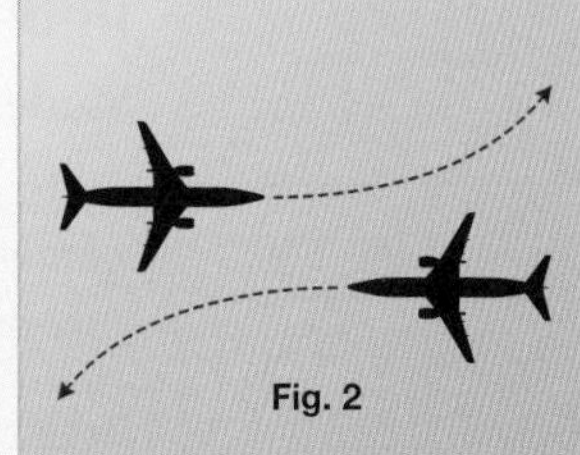
Fig. 2

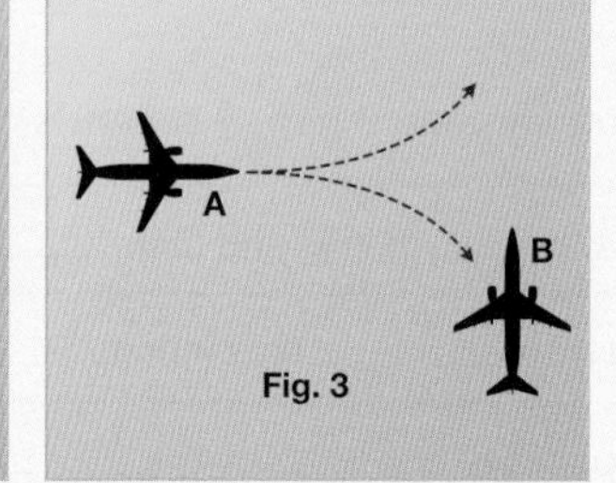

Fig. 3

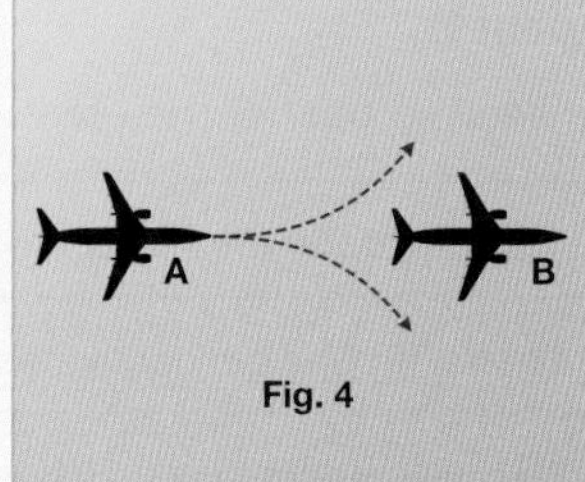

Fig. 4

The following rules apply when two aircraft approach each other at the same altitude.

1 If both aircraft are in front of each other, and see each other on their left side, both must turn right.
2 If both aircraft are in front of each other, and see each other on their right side, both must turn left.
3 If aircraft A is in front of, and to the left of, aircraft B, and aircraft B is in front of, and to the right of aircraft A, then A must turn either left or right to ensure separation with minimum change of direction.
4 If aircraft A is following behind aircraft B, then A must turn either left or right to ensure separation with minimum change of direction.

10 Disasters 2 Investigation

Task exercise 7 page 77

Student A

HYATT REGENCY

THE COLLAPSE

On the evening of July 17, 1981, in the Hyatt Regency Hotel in the USA, many people were standing and dancing on the walkways above the hotel atrium. The second and fourth-floor walkways were suspended across the atrium from the roof by rods.

Suddenly the connections which linked together the rods failed. When this happened, the fourth-floor walkway collapsed onto the second-floor walkway and then into the atrium.

114 people were killed, and over 200 injured. Millions of dollars in costs resulted from the collapse.

An investigation was set up to find out the cause of the disaster ...

9 Design 1 Inventions

Speaking exercise 5 page 69

Student B

Main design features of LED bulb (compared with standard design)

Uses 50% less power than fluorescent bulb
Has same light quality as incandescent bulb
Safer than fluorescent or incandescent bulbs because
- *made of unbreakable plastics*
- *contains no mercury*

Dimmable (i.e. can be dimmed, unlike fluorescent bulb)

3 Events 3 Sequence (2)

Task exercise 5 page 25

Student A

As soon as the rocket engine has fired, it lifts the seat with the crew member to a height of between 30 and 60 metres above the plane, and then burns out. At this point, about 0.5 seconds have passed since the moment when the crew member pulled the ejection handle.

Once the rocket engine has burnt out, and the seat is high above and clear of the plane, the drogue gun, activated by a timer, fires a heavy metal pellet vertically upwards above the seat. The pellet pulls a small parachute, called a drogue, out of the top of the seat. As soon as the drogue is out of the seat, it opens up, and begins to slow the seat down as it starts to descend.

After the drogue has opened, an altitude sensor causes the drogue to pull out the main parachute from the crew member's backpack. This happens when the crew member's altitude is between 14500 and 11500 feet above sea level. As soon as the main parachute has started to open, the seat and leg restraints are released. After this, the sudden shock of the main parachute opening pulls the crew member out of the seat, and the seat then falls away. The whole ejection operation up to this moment should take about four seconds. The crew member should then float safely to earth.

9 Design 2 Buildings

Start here exercise 1 page 70

Photo 1: Capital Gate in Abu Dhabi.

Photo 2: Hearst Tower in New York.

Photo 3: 30 St Mary Axe (also called 'Swiss Re' or 'the Gherkin'), London. Note: a gherkin is a small type of cucumber.

Answer to question 3: All three buildings use a 'diagrid' (diagonal grid) structure as the outer shell of the building.

11 Materials 3 Properites (2)

Task exercise 8 page 89

Student B

You want to spend reduced budget on:
Sensors for rowing team

Reasons:

- *training is the most important factor and sensors will help to train rowing team*
- *money spent on training more effective than money spent on equipment*
- *rowing team are below standard – need more training for gold*
- *running teams are already good – don't need so much investment in shoes*
- *Doppler lidars very expensive – not necessary – good sailors can 'read' wind*
- *Olympic committee may ban use of SpeedShark swimsuits, so swimsuits may be waste of money*

8 Projects 3 Drilling

Speaking exercise 9 page 63

Student B

FACT SHEET STAGE 1B:

Electronic guidance system

Purpose: to control the direction of the drilling along planned path

How guidance system works

1. fit electronic transmitter immediately behind drill bit
2. transmitter sends signal to locator at surface directly above drill bit
3. locator indicates exact location of drill string
4. take reading every 2 to 5 metres to check position of drill bit
5. if location incorrect, adjust direction of drill bit
6. drill bit reaches exit point and comes out of ground
7. detach transmitter from string and attach reamer

Review Unit B

Exercise 10 page 34

The photograph on page 34 shows the Falkirk Wheel in Scotland. It lifts and lowers boats from one level to another. Archimedes discovered that floating objects displace their own weight. This means that each water-filled container (with or without a boat) weighs the same amount and is therefore exactly counter-balanced. As a result, the machinery only uses 1.5 kilowatt-hours of energy – the same amount used to boil eight kettles of water. It can lift 600 tonnes of water and ships (300 tonnes in each container). The height of the upper canal is 24 metres. The wheel has an overall diameter of 35 metres with two opposing arms, each extending 15 metres from the central axle. Despite its size, it takes the giant wheel only slightly over 5 minutes to turn through 180 degrees. It is, in fact, the second largest and most advanced large-scale rotating boat lift in the world.

10 Disasters 2 Investigation

Task exercise 7 page 77

Student B

HYATT REGENCY

ORIGINAL DESIGN OF WALKWAY CONNECTORS

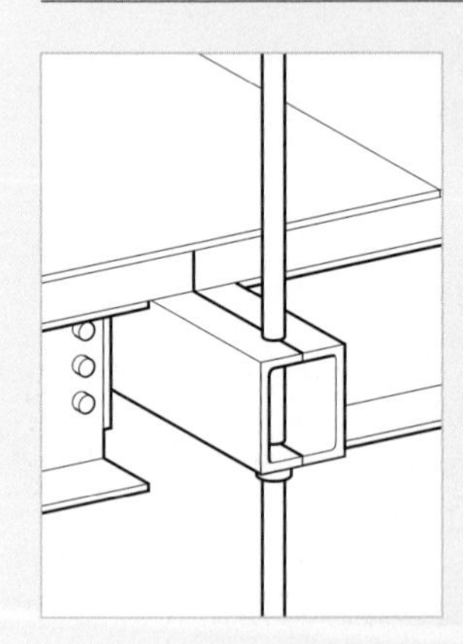

In the engineer's original design, the second and fourth floor walkways above the atrium were both supported from the roof by long vertical rods. The two walkways were also connected to each other using the same rods. However, this design was changed during construction ...

8 Projects 3 Drilling

Speaking exercise 9 page 63

Student C

FACT SHEET STAGE 2:

Reaming the pilot hole

Purpose: to ream* the pilot hole until it is wide enough to contain the product pipe

1. remove drill bit and transmitter from drill string
2. attach reamer to drill string
3. pull reamer backwards from exit point towards drill rig
4. reamer rotates to cut and widen hole
5. pump drilling fluid through string
6. this removes pieces of rock and enlarges hole more
7. reamer deposits layer of bentonite** behind it
8. bentonite ensures wall of hole is clean, hard and stable

* to *ream* a hole means to widen it

** *bentonite* is a substance which can seal and harden the walls of a tunnel

11 Materials 3 Properties (2)

Task exercise 8 page 89

Student C

You want to spend reduced budget on:
(a) SpeedShark swimsuits for swimming team, and
(b) Doppler lidar system for sailors

Reasons:

- *swimming team are excellent – SpeedShark may give many golds*
- *other teams will use them – disadvantage for us*
- *there are no reports that Olympic committee will ban SpeedShark*
- *Doppler lidar systems used in Beijing Olympics – gave edge to sailing teams*
- *no chance of golds without Doppler lidar: superior technology wins golds*
- *running teams are already good – don't need so much investment in shoes*
- *rowing teams are below standard – sensors won't help them much*

Task exercise 2 page 94

Hydrogen fuel cells

- pure hydrogen contains no carbon, and does not create CO_2 or CO
- the only emission from a hydrogen fuel cell is water
- you can drive over 320 km with a full tank of compressed hydrogen
- hydrogen is the lightest atom so it can escape from the smallest leaks
- storage, transportation and distribution are difficult: it's stored either in liquid form (at –253°C) or under extremely high pressure (about 700 bar)
- there is a risk of combustion or explosion during re-filling with hydrogen
- the fastest methods of producing hydrogen consume energy and emit highly toxic chemicals
- the cleanest way to produce hydrogen is to break up water with electricity, but many parts of the world have water shortages
- when mass-production begins, the cost of a fuel cell engine is estimated to fall to €150 per kW. (A car needs about 60 kW to run.) Industry's ultimate goal is €20–€35
- hydrogen will probably cost the driver about 2.5 euro-cents per km

Non-contact recharging

- in one system, car is driven and parked in a designated parking bay with an induction coil; driver can do shopping while car is recharging
- in another system, the induction coil is built into the road: car is recharged as it drives
- no need for cable connection; no possibility of electric shock; no need for driver to carry a mains electric plug or other equipment
- a car battery can be charged in about three hours
- an induction coil is a low-cost simple technology which will not add significantly to the cost of an electric vehicle
- it will probably cost around 2–2.5 euro-cents per km to charge a battery
- you can drive about 160 km with a fully recharged battery

Battery exchange

- car is driven into a special area like an automated car wash where the battery is exchanged for a newly-charged one
- fully automatic system: the driver doesn't have to leave the car
- the exchange takes about 3–5 minutes, less time than it takes to fill a petrol tank
- battery itself is free; driver pays a refundable deposit for the first battery, so the driver only pays for the electricity; the deposit is repaid when the battery is returned
- one company estimates that drivers will pay less than 2 euro-cents per km for the electricity in a battery-exchange system
- with a fully-charged battery, a typical saloon car could travel 160–190 km

Solar cars

- the sun is an unlimited and free resource (but it does not shine all the time)
- fitting battery electric vehicles with solar cells would increase their range beyond the normal 160 km; they would allow recharging while parked anywhere in the sun
- energy can be stored in batteries or supercapacitors for use when sun is not shining; one company has developed a supercapacitor which can give a range of 250 km after a 5-minute charge
- solar panels on roofs of cars have a size limit and may not provide enough power; the sun supplies energy of about 1 kW per m^2; the maximum size of a solar panel is less than 10 m^2, so the maximum power available to a solar car is 10 kW; however, today's car engines need about 60 kW
- panels on cars are very heavy and add to the overall weight of the car
- solar panels are expensive and can be easily damaged

9 Design 3 Sites

Speaking exercise 6 page 73

Student A

Describe buildings 1, 5, 7, 9, 11, 13 and 15 in random order to Student B without mentioning the building letter or name. Use some of the words and phrases for describing shapes in exercises 2 and 5. If B identifies a building correctly, move on to the next one. If B gets it wrong, describe the building again in a more detailed way to distinguish it from the others.

When B has identified all the buildings, switch roles and try to identify the buildings that B describes.

10 Disasters 2 Investigation

Task exercise 7 page 77

Student C

HYATT REGENCY

CHANGES IN DESIGN DURING CONSTRUCTION

After the builders started constructing the hotel, they made some changes to the design of the connections between the second and fourth floor walkways.

The builders changed the original engineer's design from a one-rod to a two-rod system to make the job easier. This doubled the load on the fourth floor walkway.

The engineering firm later said (in court) that they did not give approval in writing to the builders to make these design changes.

8 Projects 3 Drilling

Speaking exercise 9 page 63

Student D

FACT SHEET STAGE 3:

Pulling back the product pipe

Purpose: to fit the product pipe inside the reamed hole

1 remove reamer from drill string
2 attach drill string and reamer to a swivel
3 attach swivel to product pipe
4 this prevents rotating string from twisting the product pipe
5 pull product pipe along drill hole
6 pump drilling fluid along hole
7 this lubricates product pipe and allows it to pass along hole smoothly

11 Materials 3 Properites (2)

Task exercise 8 page 89

Student D

Prepare for your role as chairperson; you need to:

- *take brief notes for the minutes*
- *make sure that each person has an equal chance to speak*
- *keep order*
- *make sure the meeting doesn't run over time (allow approximately ten minutes)*

10 Disasters 2 Investigation

Task exercise 7 page 77

Student D

HYATT REGENCY

REVISED DESIGN OF WALKWAY CONNECTORS

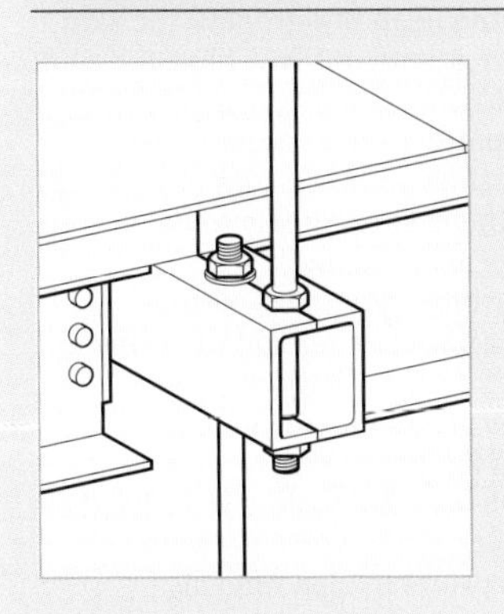

In the revised design, the fourth floor walkway was supported from the roof using one set of rods. However, the second-floor walkway was supported from the fourth-floor walkway (not directly from the roof) using a different set of rods.

5 Safety 3 Rules

Speaking exercise 6 page 41

Group B

Rules of the Air (Part 3): Phase-of-flight priority

The following rules apply if two aircraft approach each other in a different phase of flight (climbing, descending or cruising) from each other.

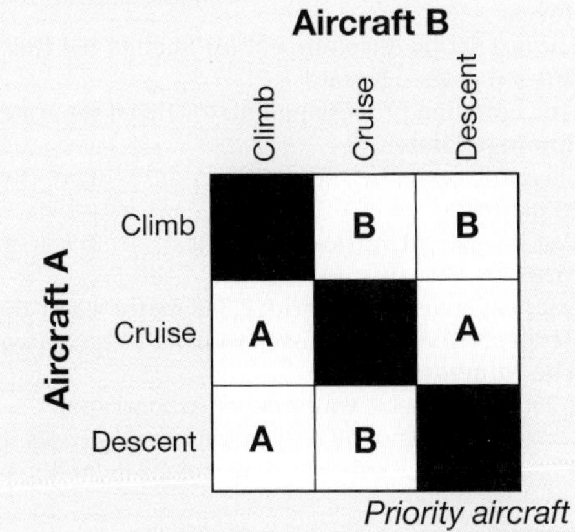

1 Cruising has priority over climbing or descending. If aircraft A is cruising, and aircraft B is climbing or descending, A has priority, and B must take evasive action.
2 Descending has priority over climbing. If aircraft A is descending, and aircraft B is climbing, A has priority, and B must take evasive action.

3 Events 3 Sequence (2)

Task exercise 5 page 25

Student B

Ejection seats look like regular seats, but they have rollers which are attached to rails in the cockpit. The rails are angled in the direction of ascent. When it is activated, the seat moves rapidly upwards along these rails.

The ejection seat system is activated when the crew member pulls the ejection handle on the seat. As soon as the crew member has pulled this handle, the explosive bolts which attach the canopy to the cockpit wall explode and break up. This allows the canopy to become detached from the cockpit, and the canopy then flies away from the plane. Immediately after this, the explosive cartridge of the catapult gun is fired.

As soon as the gun has fired, the rollers of the seat start moving upwards along the guide rails, and the seat starts to rise. At the same time, the leg restraint system is activated, and the restraints pull the crew member's legs tightly to the seat. This restrains the crew member's legs and prevents them from hitting anything while the seat is moving at speed. After this has happened, the seat leaves the guide rails and shoots out of the cockpit. As soon as the seat has cleared the cockpit, the rocket engine, which is located under the seat, fires and propels the seat up and away from the plane. All of this takes about 0.15 seconds from the moment the crew member pulls the ejection handle.

9 Design 3 Sites

Speaking exercise 6 page 73

Student B

Try to identify the buildings that A describes. When you have done this, switch roles. Describe buildings 2, 4, 6, 8, 10, 12, 14 and 16 in random order to Student A without mentioning the building letter or name. Use some of the words and phrases for describing shapes in exercises 2 and 5. If A identifies a building correctly, move on to the next one. If A gets it wrong, describe the building again in a more detailed way to distinguish it from the others.

Speed search

The world's deepest underwater post box is located 10 metres beneath the waters of Susami Bay, Japan. It's used by passing divers to send mail up to addresses on land. The post box is emptied daily by the Susami post office.

Our next big idea: a wind-charged electric car
We're well underway with our work on a second generation wind-powered car, one that you could actually drive to work or go shopping in ... we're building it right now. Actually it's an electric car, charged from the wind – but not just any electric car, one to smash the stereotype. An out-and-out sports car. Capable of accelerating from 0 to 100 kph faster than a V12 Ferrari, able to top 160 kph for sure – and do 240 km on one 'tank'. All with zero emissions. We're making this car with technology that's available in the world today and throwing down the gauntlet to the big car companies. Our message to them is: "If we can do it, and we're just a tiny electricity company – why can't you?"

Planes just metres apart in near miss
Controllers insist there was no trace of light aircraft
A mid-air collision close to Malta's airport in April last year between a passenger jet and a light aircraft was avoided when the planes were just 180 metres apart, a report said.
Only quick thinking by the pilot of an Air Malta Boeing 737, making its final approach at the end of a flight from Rome with 80 passengers on board, managed to prevent a potential disaster over a populated area two nautical miles away from Malta International Airport.
The Italian pilot of the light aircraft claimed that he could not find a suitable place to land, and therefore had to make an emergency landing at MIA.
The light aircraft was not seen by air traffic control, and therefore entered Maltese controlled airspace undetected.
Meanwhile, as the Boeing emerged from cloud at approximately 1,300 feet, the Air Malta pilot was shocked to see his aircraft on a collision course with the light aircraft.
The Boeing pilot took evasive action and made a right turn, descending to 800 feet, followed by a climb. Then the Boeing descended further, in order to avoid hitting the light aircraft with its left wing.

The heaviest building ever moved intact is the Fu Gang Building at West Bank Road Wuzhou, in the Guangxi Province of China. It was successfully relocated by the Guangzhou Luban Corporation on November 10, 2004. The building weighs 15,140.4 metric tonnes and is 34 m tall. The building was moved 35.62 m horizontally, and it took eleven days to complete the relocation.

- *Sarsat* system – developed by US, Canada, France
- *Cospas* system – developed by Soviet Union (Russia)
- 1979 – four nations combined into a single *Cospas-Sarsat* system
- 1982 – first *Cospas-Sarsat* satellite launched
- 1984 – system fully operational
- Today – system joined by at least 25 more nations

Some facts about fingerprints
Every person has a different set of fingerprints.
Fingerprints are formed in the foetus before birth.
They are caused by ridges in the flesh underneath the skin.
Even identical twins (who have identical DNA) have different fingerprints.
More than 60% of fingerprint patterns are 'loops'.
The same ridges cover your palms and the soles of your feet.
Two murderers in 1905 were the first criminals to be convicted by means of fingerprint evidence.
You can't change or alter your fingerprints by cutting, burning or scraping them. The patterns remain the same as new skin is developed.
The earliest dated use of fingerprints for transactions were made about 4,000 years ago in Egypt.
In ancient Babylon, fingerprints on clay tablets were used as signatures on contracts.
The Chinese used inked fingerprints to sign official documents.
In fourteenth-century Persia, many government papers had fingerprints on them.

Some facts about the launch abort system

Maximum altitude for system to operate	100,000 m
Length of LAS (detached from crew capsule)	13.36 m
Thrust of LAS abort engine	over 180,000 kg
Top speed of LAS ejection	725 kph
Time taken to reach top speed	3 seconds
Length of crew capsule (detached from LAS)	2.64 m
Pressure experienced by crew during ejection	11 Gs

The Perdido Spar offshore oil rig
Features
- Largest scope development for Shell in the Gulf of Mexico
- Rugged seabed terrain
- One common processing hub for three separate oil fields

Technology Firsts
- Oil and gas will be separated on the seabed, then pumped to platform
- Wet tree direct vertical access wells from a spar

Records
- Deepest spar in the world: 2,383 metre water depth
- Deepest producing sub-sea well: 2,934 metre water depth

By The Numbers
- 2,280–3,050 metre water depth around spar
- 130,000 barrels of oil equivalent per day capacity
- 45,360 tonnes total operating weight including vertical tension
- 170 metres in length, 36 metres in diameter
- 35 wells (22 direct vertical access sub-sea, 13 remote sub-sea)
- 322 kilometres south of Houston, Texas

Milestones
- 1996: lease sale
- 2002: first discovery of oil
- 2007: drilling begins
- 2008: spar installed on site
- 2009: topsides installed

Introduction
There are more than one trillion tonnes of coal in the world. But coal emits harmful pollutants when it is burned. The purpose of clean coal technology (CCT) is to reduce these harmful emissions. There are four main CCT processes.
1 Cleaning coal before burning
The coal is ground into smaller pieces and passed through a special fluid inside a gravity separator. The fluid causes the coal to float, and allows the impurities to sink. The impurities are then removed and the cleaned coal is pulverised (ground into dust).
2 Desulphurisation
Sulphur is removed from the flue gas (the gas emitted from burning coal) by spraying a mixture of limestone and water over the gas. The fluid reacts with the SO_2 in the gas to form gypsum, which is then removed.
3 Removal of particulates
Particulates (small polluting particles) are removed from the flue gas by electrostatic precipitators. An electrical field is generated in the particles, which are then attracted by collection plates and removed via hoppers.
4 Carbon capture and storage (CCS)
Carbon dioxide emissions are captured and stored deep underground to prevent the greenhouse gas from entering the atmosphere. The CO_2 can be pumped (1) into disused coal fields, displacing methane which can be used as fuel, (2) into saline aquifers (water channels), where it can be stored safely, or (3) into oil fields, which helps maintain pressure, making extraction easier.
Conclusion
There is enough coal in the world to last for 150–200 years. Provided that the processes detailed above are applied, coal can be transformed into a low-carbon-emission clean energy source

The smallest purpose-built cinema still in operation is the Cinema dei Piccoli, built in 1934 in the park of Villa Borghese, Rome, Italy, which today covers an area of 71.52 m². Originally called the Topolino Cinema (after Mickey Mouse), the cinema used a Path-Baby 9.5 mm movie projector, bed sheets for the screen and played 78s for background music. Restored in 1991, the cinema has 63 seats, a 5x2.5 m screen, stereo sound and air-conditioning.

Electrical Design Manager (Ref: ED-207)
Our Client wishes to recruit an experienced Electrical Design Manager with experience in designing electrical power-generating equipment.
The job involves the development of existing engineering equipment and the generation of new designs.
Experience in the following areas is preferred:
- generator sets and their control systems
- LV, MV and HV systems
- alternators, transformers and switchgear
- use of AutoCAD software

Duties will include:
- equipment design and development
- generation of electrical schematic drawings
- product support documentation
- some travel in Europe and the Middle East

You will have a minimum of a Higher National Diploma (HND) in Electrical Engineering or an equivalent qualification. The role requires a self-motivated individual capable of working with the minimum of supervision.
In return they are offering a salary of €47–50K plus a bonus of between 20% and 50% depending upon targets. Benefits include pension, life insurance, and medical care.

Grande Dixence, on the river Dixence in Switzerland, is the highest concrete dam in the world. It was built between 1953 and 1961 to a height of 285 m (935 ft), with a crest length of 700 m, using 5,960,000 m³ of concrete.

PLASTIC	PROPERTIES	SPORT
Carbon fibre (= plastic reinforced with carbon fibre)	Strength, durability, abrasion resistance, lightweight, impact resistance, shatter resistance, stiffness in arrows, boats, bicycle frames, tennis racquets	Archery, Canoeing, Cycling, Kayaking, Rowing, Tennis
EPP (= expanded polypropylene) foam	Impact absorbency for helmet liners	Canoeing, Kayaking, Rowing
Fibreglass (= plastic reinforced with glass fibre)	Lightweight, durability flexibility, impact resistance, abrasion resistance, shatter resistance in arrows, boats, oars, gymnastics bars, discs, and batons	Archery, Canoeing, Kayaking, Field Hockey, Gymnastics, Rowing, Sailing, Track & Field
Spandex	Flexibility, moisture absorbency, aerodynamic, comfort for clothing	Cycling, Diving, Swimming, Synchronized Swimming, Water Polo, Wrestling
Neoprene	Impact absorbency, slip-resistance, in wrist guards and weightlifting shoes	Gymnastics, Weightlifting
Nylon	Durability, lightweight, elasticity, strength, water-resistance in knee pads, bicycle tyres, harnesses and reins, fencing clothing, sailing ropes, tennis racquet strings,	Volleyball, Cycling, Equestrian, Fencing, Sailing, Tennis, Water Polo, Weightlifting
Polycarbonate	Shatter resistance, abrasion resistance, optical clarity, impact resistance, durability in protective goggles and leg guards	Cycling, Diving, Field Hockey, Gymnastics, Shooting, Swimming, Synchronised Swimming, Tennis
Polyester	Durability, elasticity, strength, water resistance in volleyball nets, sailing ropes, and tennis racquet strings	Volleyball, Sailing, Tennis

Audio script

Unit 1 Systems

02

A dramatic air-sea rescue took place at 11 a.m. this morning in the Indian Ocean. Two sailors were pulled up from the sea into a helicopter using a powerful winch, in very rough seas and a high wind.

The sailors were in a small boat, the *Tiger*, about 77 kilometres from land. Suddenly their boat struck an object, and it sank almost immediately.

The sailors wanted to send an emergency signal by radio, but the boat went down too quickly, and the radio sank with the boat. So the men did the best thing to save their lives – they inflated their life raft and jumped in. They were already wearing their life jackets, of course. But their problem was how to call for help 77 kilometres from land.

Fortunately, the boat had a free-floating beacon, which activated when the boat sank to four metres. The beacon detached itself from the boat, floated to the surface of the sea and switched on automatically. Then it transmitted a signal to the rescue satellites.

The rescue team received the emergency signal, and raced in their helicopter to the two men in the life raft.

When the helicopter came close to the life raft, the sailors fired two flares. The pilot saw the flares and brought the helicopter over the life raft. Then the sailors were winched up.

The whole rescue operation, from the moment the *Tiger* sank, took only 90 minutes, thanks to the automatic beacon and the satellite system it was linked to.

03 optional listening

04

A: So how does the rescue service work? What happens after your plane crashes, or your ship starts sinking?

B: Well the first thing you do, if you're a survivor of a plane crash or a sinking ship is to activate your personal emergency beacon manually.

A: OK, but what happens if I can't locate my personal beacon, or it sinks with my ship?

B: Well, most planes and ships today are fitted with automatic beacons. So when the plane hits the ground or the ship starts to sink, the beacon detaches itself and activates itself automatically.

A: Right, so let's assume my beacon is activated. What happens next?

B: Well, the beacon then transmits a radio signal, and one or more satellites receive the signal and ...

A: ... and the satellites then send the signal to the rescue team?

B: No, they can't do that, not directly. First they send the signal to their ground station. And the ground station then processes the signal, in other words it converts or changes the signal into useful data.

A: Right, I see. And what happens next?

B: Well, the ground station then passes this data on to a national centre. And the national centre then forwards the data to the rescue centre which is nearest to the location of the crashed plane or sinking ship.

A: I see. And the rescue centre sends out the rescue team?

B: Yes, that's right. First it locates the beacon, in other words, it finds out its exact position and marks it on a map. Then it sends out the rescue team.

A: And the rescue team carries out the rescue?

B: That's right. The team searches for the survivors, finds them, winches them into the helicopter and then takes them back to the rescue centre or straight to hospital.

Unit 2 Processes

05

1 Engineers in Germany have constructed Europe's first plastic road bridge. The bridge comes without nails and screws. Instead, the complete deck of the bridge is made of a composite plastic called fibreglass-reinforced polymer. The whole deck was glued onto two steel columns. The bridge is 27 metres long and it can be used for cars, lorries and pedestrians, just like an ordinary bridge.

2 A major aircraft manufacturer has announced that it has designed a new type of aircraft with a plastic body. The company will start to manufacture the plane at the end of next year. More than fifty percent of the fuselage, or body, of the plane will be made of a composite plastic material. This is a big step towards making a one hundred percent plastic fuselage.

3 Plastics packaging material is going to be smart, even intelligent, in the near future. A futurologist working for a major plastics manufacturing company has made a prediction for the next ten to fifteen years. He expects that we will see plastic packaging that can detect changes and give information to the customer. For example, the plastic packaging around food will contain very thin electronic chips. These chips will be able to detect changes in temperature, or changes in the freshness of the food. The plastic food package will then communicate this information to the customer.

4 A futurologist who works for a big international plastics manufacturing company has made an important prediction. He believes that by the year 2035 most cars and other land vehicles will have bodies made completely of plastic composites. However, he says, it is unlikely that cars, including their engines, will be completely made of plastics by that date.

5 Plastic ice is spreading over Scandinavia! Sweden now has more than nine ice skating rinks which are made ... not of ice ... but of plastics. Engineers in a major plastics company have designed and manufactured a new kind of plastic composite which feels just like ice. It has exactly the same friction and slipperiness as real ice, and the skates cut into the plastic to exactly the same depth as in real ice.

06

Good morning everyone, and thanks for coming. This is the fourth short talk in our series of talks about the plastics industry. Last week we looked at the process of *injection moulding*. Today, I'm going to explain how *extrusion blow moulding* works.

Extrusion blow moulding is a method of making a hollow shape out of a thermoplastic. This shaping method is very useful for making things such as plastic bottles, petrol containers, jerry cans and so on.

As its name suggests, extrusion blow moulding consists of two separate processes.

The first one is the *extrusion* process. This is very similar to the injection moulding process we looked at last week. During extrusion, solid pellets, or small pieces, of plastic are heated, melted, pushed along a cylinder and extruded, or pushed out, into a mould.

The second process, *blow moulding*, takes place inside the mould, where compressed air blows into the centre of the molten plastic and expands it into a hollow shape such as a bottle.

Let's look at the first process, the extrusion of molten plastic into the mould.

As you can see in Figure 1, there is an extruder at the top left of the diagram. This operates like the injection moulding cylinder we saw last week.

As I'm sure you will remember, first of all, pellets of raw plastic are fed from a hopper into a large horizontal cylinder. Inside this cylinder, a large screw rotates. This rotation pushes the cold polymer pellets along the cylinder towards the right. There are heaters all along the sides of the cylinder. These heaters heat up the polymer pellets and melt them. The screw continues to push the soft, melted polymer along the cylinder.

As can be seen in Figure 1, there is a ninety-degree angle at the right-hand end of the cylinder. This angle, or bend, is inside the die in the top right-hand section of Figure 1. Now the molten plastic flows downwards through the die, and is extruded, or pushed out, into the mould.

So that's the end of the first part of my talk, about the *extrusion process*. Let's turn now to the second part, which is about the *blow moulding* process.

Blow-moulding consists of three stages. I will now describe each of the three stages in turn.

Let's look at the first stage. As Figure 1 shows, the hot, soft plastic is extruded down between the two halves of the open mould. The plastic is in the form of a long, hollow tube, called a parison. Then, as Figure 2 illustrates, the two halves of the mould close. Now the parison is inside the mould.

The second stage is illustrated in Figure 2, as well. In this stage, compressed air is blown through the nozzle into the molten polymer parison. The air inflates the parison, and as a result, the soft plastic expands to fit the shape of the mould. The plastic is cooled by the cold surfaces of the mould. This sudden cooling causes it to harden quickly in the shape of, in this case, a bottle.

The third and final stage is shown in Figure 3. Here, after a cooling period, the two halves of the mould open, and the bottle is ejected from the mould onto a conveyor belt.

Unit 3 Events

07

[R = Reporter; TC = Technology Correspondent]

R: This is the six o'clock news for today, the 14th of April, 2020. First, the news in brief. The new Ares moon rocket has failed to launch. The rocket has crashed into the Indian Ocean. The crew capsule containing the six astronauts has landed safely in the ocean. Rescue helicopters have taken the astronauts to hospital, where they are recovering. The director of the space program has resigned.

Now the news in detail. Six astronauts escaped death early this morning when their Orion crew capsule detached itself safely from their Ares space rocket. The Ares rocket was launched at 5.05 this morning but after only a few seconds it was obvious that something was seriously wrong.

At first the rocket flew straight upwards, but then it turned and moved almost horizontally before starting to fall back towards Earth. The rocket then crashed into the ocean and disappeared from sight.

For more details about the ejection system on the Ares rocket, we can now turn to our technology correspondent, Jeff Walker, who is at the recovery site.

TC: Luckily for the crew, the Ares rocket was fitted with the new Launch Abort System. If something goes wrong within the first 100,000 metres of the rocket's ascent, the Launch Abort System, or LAS, is activated and carries the crew capsule away to safety, just like the ejection seat system in an aeroplane.

The LAS is at the nose, or tip, of the rocket, and contains an abort engine and a supply of solid fuel. It is attached to the crew capsule.

As soon as the launch failed, and the Ares rocket changed course, the LAS was automatically activated. The abort engine fired with a massive 180,000 kilograms of thrust, and the LAS, attached to the Orion crew capsule, was ejected and shot upwards at high speed and away from the rocket. It reached a speed of 725 kph in less than three seconds.

When the solid fuel burnt out, the crew capsule detached itself from the LAS. Three sets of parachutes then opened up, and the crew capsule floated down and landed safely in the ocean.

Helicopters reached the capsule within a few minutes, and took the six astronauts to hospital at their base to recover. All six are well and in good spirits. Sarah.

R: Thanks, Jeff.

08

... as the Challenger disaster showed.

So now I'd like to tell you about our dream for the future of space travel, a future that will make safety a reality for our astronauts. I'm talking about an ejection system for astronauts!

If you look at the first slide, this describes the reality we have today. We have ejection seats today, but only for aircraft.

If an aircraft fails in some way, the pilot will activate the ejection system. And if the pilot activates the ejection system, the system will immediately eject the ejection seat, with the pilot, from the plane.

So let's turn to the second slide, which describes our dream for the future. We don't have ejection capsules yet for spacecraft, but one day soon, with the LAS system, our new invention, we will.

Let me describe what would happen if we had an LAS system in our spacecraft today.

If a spacecraft failed in some way, the computer would activate the LAS system. And if the computer activated the LAS system, the system would eject the capsule, with the crew, from the spacecraft.

Now, as Figure 1 here illustrates, the ejection system for aircraft ...

09

I'd like to spend a few minutes now describing the main parts of the LAS, or Launch Abort System, of the Ares space rocket.

As you can see in Figure 1, on the left-hand side of the illustration, the Launch Abort System is right at the nose, or tip, of the rocket. It's shaped like a dart, or a pawn on a chessboard. The job of the LAS is to lift the astronauts to safety if something goes wrong during the rocket launch.

Now let's look at the LAS in more detail. As can be seen in Figure 2, the cone-shaped structure attached to the bottom of the LAS is the Orion crew capsule. This is where the four to six astronauts live and work.

As you can see, a protective cover surrounds the Orion crew capsule and shields it from the hot engine exhaust.

Inside the LAS, the abort engine provides the power to lift the complete LAS and crew capsule at high speed away from the falling rocket. The exhaust from this engine escapes through nozzles in front of the engine.

Below the abort engine in the LAS you can see the solid propellant. This is the fuel for the abort engine. All two thousand one hundred kilos of fuel in the abort engine burn in less than three seconds.

Right at the tip, or nose, of the LAS, as Figure 3 shows, there is a ring or circle of four small nozzles. These are the nozzles of the attitude-control engine. When these nozzles fire, they spin the LAS around to stabilise it and orient it.

Just above the crew capsule, as Figure 4 illustrates, there is a third engine, called the jettison engine. When this engine fires, it pushes the crew capsule away from the LAS.

And finally, in Figure 5, you can see the parachutes. When the capsule reaches a safe altitude, the parachutes open up and bring the crew back to Earth.

10 optional listening

Unit 4 Careers

11

[I = Interviewer; HF = Hans Fischer]

I: I'm talking to Hans Fischer today. Hans is a young German engineer who is currently working at Farmakon International, a large German pharmaceutical company. He also happens to be an inventor in his spare time. Good morning, Hans.

HF: Good morning.

I: You've become famous because of your latest invention. But you're not really an inventor, are you? Or at least, inventor is not your job designation.

HF: That's correct. My actual job title is Robotics Engineer.

I: So, what is your job description? What do you actually do every day?

HF: I design and build robots for the packing lines. Then I test them. I work with robots every day.

I: And how long have you worked at Farmakon?

HF: Well, I've worked here as an engineer for the last two years. Before that I was at university for four years, and before that I worked at Farmakon for six years. So I've worked at Farmakon for eight years altogether.

I: I see. So how old were you when you first started work at Farmakon?

HF: I was sixteen. I left secondary school and then I joined Farmakon as an apprentice technician.

I: And how long was your apprenticeship?

HF: It lasted for three years. During my apprenticeship, I studied part-time at vocational school.

I: And then after your apprenticeship ended, you became a technician at Farmakon. Is that right?

HF: Yes, that's correct. After my apprenticeship ended, I worked here as a technician for three more years while I studied part-time for university entrance. Then I went to study full-time at the University of Applied Sciences in Munich.

I: What course did you do there?

HF: A Bachelor of Engineering degree in Mechatronics. My main subject was Robotics. I got my degree two years ago.

I: Congratulations!

HF: Thanks.

I: And then after university did you come straight back to work at Farmakon?

HF: Yes, that's right. That's when I became a Robotics Engineer here.

I: And did you have your idea for an invention soon after returning to Farmakon, after your degree?

HF: Yes, that's right.

I: So tell me about your invention. What is it?

HF: Well, I've designed a robot that can work safely with a human worker.

I: That's very interesting. Can you use it at Farmakon?

HF: Oh yes, I'm piloting it now on one of the packing lines. I'm trying it out and testing it.

I: And what's your next step? What do you intend to do next?

HF: Well, next month, if the pilot is successful, I'm going to build another robot for a different packing line.

I: Excellent. And what are your plans for your career from now on?

HF: I'm doing a Masters degree next year. It's a distance-learning course. I've already been given a place on the course.

I: That's great. Well, Hans Fischer, it's been very interesting hearing about your achievements and your plans. Good luck with your future career. Thank you very much.

12 optional listening

13

[I = Interviewer; RG = Reme Gomez]

I: ... So, Ms Gomez, can I check a few details from your CV? When did you complete your school education?

RG: I left school just over seven years ago.

I: And now you're a technician?

RG: Yes, that's right. I'm a junior technician at MultiPlastics.

I: I see. And how long have you worked there?

RG: Erm, I went there straight from school so I've been there for seven years. I started as an apprentice.

I: How long did your apprenticeship last?

RG: For three years.

I: And after your apprenticeship ended, they promoted you to junior technician. Is that correct?

RG: Yes, that's right. I've been a junior technician there for the last four years.

I: And you're currently studying part-time for a technician's diploma, are you?

RG: Yes, I am. I go to the technical college twice a week.

I: And how long have you been a part-time student?

RG: For two years now. I started the diploma course just over two years ago.

I: And you already have your Certificate of Technical Competence, of course?

RG: Yes, I studied part-time for my certificate while I was an apprentice. My employer gave me time off work, to help me. I got my certificate five years ago.

I: Right, thank you, I think the details are clear now. So, Ms Gomez, why would you like to work with us?

RG: Well, you're one of the leading companies in your field. I'd like to broaden my experience.

I: And why should we offer you this job? What technical skills do you have?

RG: Well, my benchwork is very accurate. And I've learned CADCAM.

I: And what about your personal skills?

RG: Well, I work hard, I'm punctual and my present employer says I'm very reliable.

I: What about interpersonal skills? Do you like working with others?

RG: Yes, I'm willing to learn and I think I'm a good team worker.

I: Good. So now, do you have any questions to ask us?

RG: Yes, I would like to know a bit more about the salary and benefits ... that come with the job.

Unit 5 Safety

14

[T = Tom; M = Max]

T: Tom Redman.

M: Hi Tom, it's Max here.

T: Max! Good to hear from you. How are you?

M: Fine, thanks. How are things with you?

T: Great, thanks.

M: I hear your research project is going well.

T: Yes, I think we're making progress.

M: What is it you're looking into?

T: We're trying to find out whether car safety systems make people drive less safely.

M: That sounds very interesting.

T: Thanks. So how about your work? How is your latest product design coming along? Your last one was a big success.

M: Yes, it was very popular with the customers. And the new product is looking very good. Very good indeed. And that brings me to my reason for calling you.

T: Mmm?

M: I'm planning to hold a brainstorming session next week with the rest of the design team. I thought that your ideas would be very useful. Would you have time to join us?

T: Of course. When are you meeting?
M: Hopefully next Tuesday at ten. If you're free then.
T: Yes, I think that's OK for me.
M: Great. See you next Tuesday then. I'll send you an email to confirm it.
T: Good.
M: And if you can join us for lunch afterwards, we can have a chat about holiday plans.
T: I'm looking forward to it. Bye.

15

[M = Max; T = Tom; M = Man; W = Woman]
M: We need to think about a new warning system to help drivers keep in their lane while driving. Any ideas?
M: Well, I think that we should have infra-red sensors under the bumpers. They can monitor the lane markings.
W: Or we could have cameras inside the windscreen. They can see further ahead.
M: Yes, I agree, you have a point there, cameras are probably better. They can detect the lane markings, then the controller can give a warning. For instance, you could have a flashing light or another visible warning. Or you could have an audible warning like a beeping sound.
W: I don't think that's a good idea. Beeps and lights are too annoying and distracting for the driver.
M: So why don't we have a SatNav voice telling the driver 'You're crossing into the next lane'?
W: No, drivers wouldn't like that. I mean, it would sound like an angry school teacher. Or your wife or husband criticising you. We shouldn't use a SatNav voice. By the way, I had a very bad experience with a SatNav last week …
M: Anyway, let's keep to our main discussion. If we don't have a voice, a beep, or a flashing light, what do you suggest?
W: Well, I think we should use normal feedback signals, in other words, signals from the real world.
M: What do you mean?
W: Well, for example, if you go too fast, or leave your lane unintentionally, the car could make you feel a little fear, let's say, for example, by tightening the seat belt a little.
M: That might be too frightening and make the driver over-react. It could make him take too much corrective action and press the brakes too hard. Alternatively, he might counter-steer too much, you know, steer too much in the opposite direction.
W: You have a point there. So let's make the feedback more gentle. If the car crosses the line, the warning system makes the steering wheel vibrate a little. Just enough vibration to make the driver feel something is wrong.
M: Yes, I like that.
M: That sounds good. I think we're on to something here. What do you think of all this, Tom? You've done research into how people drive when the car is full of safety devices.
T: I think the vibrating steering wheel is a good idea. You shouldn't have a system that takes control of the driving from the driver, for example by pressing the brakes automatically. That makes people drive more dangerously, because they think that the car is completely safe. Warnings, like the vibrating steering wheel, are much better. They make the driver stay alert and responsible for his own safety.

16 optional listening

17

A: You should check your car brake system at least once a year. And while you're driving, you should notice anything unusual with your brakes.
B: What sort of thing should I notice?
A: Well, if the brake pedal feels soft or spongy, or if you have to pump it up and down to stop the car, you must check the brakes.
B: What could be wrong?
A: Well, there could be air in the brake lines or air in the master cylinder. Or the brake fluid level could be too low.
B: What do you mean, air in the brake lines?
A: Well, just look at the diagram here. You can see the disc. You can also see the brake pads, and the calliper. The disc is that circular part mounted on the hub behind the wheel of the car. Above the disc in the diagram is the calliper. The calliper, shown in green in the diagram, fits over the top of the disc. On each end of the calliper are the brake pads, marked in dark blue. These are made of a softer material than the disc to prevent the disc from being scratched or damaged.
So, when you put your foot on the brake pedal and press it down, the brake pads are squeezed together onto the disc. The calliper and the pads are just like your finger and thumb holding a plate. If the pressure is hard enough, the pads will stop the disc from moving.
B: How is the energy from the brake pedal passed to the calliper?
A: Well, if you look again at the diagram, you can see a thin pipe running from the master cylinder to the piston in the calliper. This pipe, called a 'line', contains a hydraulic fluid, which is a special type of oil. It's shown in light blue in the diagram.
So, when you press the brake pedal with your foot, a piston compresses the brake fluid in the master cylinder. This causes the piston in the calliper to push the brake pads onto the disc. When the pads squeeze the disc tightly, the disc slows down and stops.

Unit 6 Planning

18

[B = Ben; J = Jeff; D = Danielle]
B: Morning, Jeff. Morning, Danielle.
J: Morning.
D: Good morning.
B: Right, let's get started. I know you are both fully aware that there are global targets for reducing carbon emissions, cutting overall energy consumption, and increasing the use of renewable energy.
J: Yes.
B: So first let's just remind ourselves what the targets are. In fact there are *two* global targets for reducing emissions. And by the way, of course there are other greenhouse gases, such as methane – C H four – and nitrous oxide – N two O – which are bad for the environment, but we're mainly concerned with carbon dioxide – C O two – in our company, since that's the only gas we can control. Anyway, the long-term deadline for a fifty percent reduction is the year 2050 …

19

[B = Ben; J = Jeff; D = Danielle]
B: … but we're mainly concerned with carbon dioxide – C O two – in our company, since that's the only gas we can control. Anyway, the *long-term* deadline for a fifty percent reduction is the year 2050. The world is going to have to make a fifty percent reduction in emissions by that date. But the *urgent medium-term* deadline and the one that concerns us most, right now, is the year 2020. By that date the world is going to have to reduce its emissions by twenty percent.
D: Mhm.
J: Right.
B: And that means that our company, being the major energy company in the region, will have to do something similar. So let's have your ideas, please. How can we meet the twenty percent target by the year 2020? Jeff, you've done some thinking on this, so would you like to kick off?
J: Certainly, Ben. Well, my view is that we won't be able to meet the target unless we switch from normal coal-burning to CCS. What I mean by that is, we'll have to convert some of our coal-fired energy production to CCS, you know, carbon

capture and storage.

D: We'll probably be able to convert about half of our coal-burning power plants.

B: I don't agree, Danielle. CCS technology is still quite a new technology, and untested. I think a lower percentage would be more realistic.

J: Yes, I agree. Our team looked into it and decided on thirty percent.

D: OK, I can go along with that.

B: Good. So we'll agree to convert thirty percent of our coal-fired power plants to CCS. What should the deadline be, Jeff?

J: We're going to have to do it by 2015 at the latest.

B: I'm in complete agreement with you.

D: That's fine by me. But of course that won't be enough to meet the target alone.

B: You're right, Danielle. You have some ideas about our transport fleet, I think. Would you like to bring us up to speed on that?

D: By all means, Ben. At some point we'll have to replace the carbon fuel in our nationwide transport fleet – diesel oil, petrol and so on – with bio fuel.

B: That's right. So what deadline do you think we should fix for that?

D: Well, I think we'll have to convert at least ten percent of the fleet to bio fuel as quickly as possible. I think our deadline for that should be the end of 2014.

J: I'm not sure about that deadline. It's quite tight. I'm sure we won't be able to meet it. I think 2016 would be more realistic.

B: I would disagree with you there, Jeff. Every time we replace an old vehicle, we can buy one that uses bio fuel. We already have a lot of old vehicles in the fleet.

J: You have a good point there, Ben.

B: I think we're going to need to switch to bio fuels as quickly as possible. Ten per cent of the fleet by 2014 sounds right. Jeff?

J: Yes, I agree with you.

B: Good. Let's move on. Let's consider our energy consumption, as a company. We need targets for switching part of our energy supply to renewables.

D: You're absolutely right. By renewables, we're talking about wind power, solar power?

B: Yes, and hydro-electric, waves and bio fuels.

D: Well, obviously, we're going to need to get around fifteen percent of our energy from renewables.

J: I can't go along with that. I think we're going to have to increase that to at least twenty percent. I don't think fifteen percent would be enough.

B: Yes, I agree. And we'll probably need to achieve the twenty percent reduction a couple of years before the deadline, in other words by the end of 2018. What do you think?

J: Yes, I think that's correct.

D: I'm happy with that.

B: Good. So finally, and very briefly, let's look at our overall energy consumption. We're going to have to reduce that, too, probably by twenty percent, and, I think, by no later than the end of 2016. Agreed?

D: Agreed. Twenty percent by 2016.

J: That sounds about right.

B: Good. So let's get all this down on paper.

Unit 7 Reports

20 optional listening

21

A: I'm conducting an investigation into the recent security breakdown at the airport. I need to ask you some questions.

B: Fair enough.

A: You were the official on duty at security check-point B between 2 and 4 pm on the 18th of this month, is that correct?

B: Yes, that's right.

A: Good. So could you tell me exactly what happened when the passenger walked through the metal detector?

B: He walked through and the detector sounded.

A: What did you do?

B: I told him to step back, and then I ordered him to walk through again.

A: Are you sure you instructed him to walk through again?

B: Er, yes, I am. I told him to take his money out of his pockets. I told him to put the money on a tray, and then I ordered him to walk through again.'

A: Give me your exact words. What did you say to him? Actual words, please.

B: I said, 'Put your money on the tray. Now walk through again, please.'

A: And then what happened?

B: This time the metal detector didn't sound, so I told him to go on.

A: What were your exact words?

B: No words. I just waved him through.

A: What happened next?

B: My supervisor asked me what had happened and I told him that the passenger wasn't carrying any metal.

A: What were your exact words?

B: I said, 'He isn't carrying any metal.'

A: OK. Now what if I told you that the passenger was in fact carrying a knife?

B: What? No, it's not possible!

A: And what if I told you that the 'passenger' was in fact a security inspector?

B: Oh.

22

[A = Adam; B = Bob]

A: Good morning, Bob. I'd like to have a chat about your security project. You've been looking into the different security methods, I believe, is that right?

B: Yes, I'm trying to decide which security system would be best for our offices.

A: Right. So how are you getting on?

B: Fine. I'm making good progress.

A: Good. Have you made any decisions so far?

B: Well, I've looked into passwords, pin numbers, and voice recognition.

A: Aha.

B: And I've decided not to recommend any of those, for various reasons.

A: Right. So what are you looking at now?

B: Well right now I'm looking into different methods of fingerprint scanning.

A: I didn't know there were different methods.

B: Yes, there's optical scanning, which basically takes a photo of the finger, and there's something called capacitive scanning, which uses electrical current and a capacitor.

A: OK, and what have you come up with.

B: Well, I've decided against optical scanning, because it's too easy to forge a fingerprint. I mean, you could place a photograph of a finger onto the scanning plate instead of an actual finger.

A: Oh dear, yes, I see what you mean. So what about the other one, capacitive scanning?

B: I'm looking into that at this very moment. It looks a bit more secure because it measures the actual ridges, not just a picture of ridges.
A: Good. Oh, by the way, what about this new iris scanning technology. A method of scanning the eye. Have you looked into that yet.
B: No, not yet. That's a big research area, so I'm planning to have a look at that next week.
A: OK, I'm glad it's going well. I'll catch up with you next week.
B: Cheers.

23 optional listening

Unit 8 Projects

24

1 This is the News at Ten on Monday, June the 2nd, 2008. Good evening. The Perdido Spar has been towed to its site in the Gulf of Mexico. The spar, which is expected to be the deepest oil spar in the world, weighs about 45,000 tonnes, equivalent to about 10,000 motor vehicles.
2 This is the Nine o'clock News for today, the 14th of August, 2008. Good evening. The Perdido Spar has been secured to the seabed. A total of nine polyester mooring lines were used to moor the world's deepest spar, averaging more than three kilometres in length. It took 13 days to complete the job.
3 June 11th, 2009. This is the Early Evening News with Don Gomez. The topside has been fitted to the top of the Perdido Spar. The topside, which includes the drilling platform and accommodation block, was attached to the spar in calm seas early this morning. The spar itself is 170 metres long and 36 metres in diameter. The depth of the seabed below the spar platform is 2383 metres, which makes it the deepest spar in the world.
4 Good evening, this is the News at Ten for today, November 10th, 2010. The first oil well under the Perdido Spar has been completed. The well, under the world's deepest spar, has a depth of almost 2500 metres below the seabed, making the total depth of the well about 4800 metres below the topside, or spar platform.
5 This is the World Today for March the 12th, 2013. The headlines. A huge pipeline network has been laid under the Perdido Spar, totalling 300 kilometres in length.
6 July 28th, 2014. This is Global News Today. Good morning. Five risers and a pumping station have been built below the world's deepest spar. Far below the Perdido Spar, the pumping station, which is the size of a large truck, will separate the oil from the gas as it flows from the trees, or wellheads. It will then pump both fuels from the trees on the seabed, up the risers to the topside, using one thousand one hundred and twenty-kilowatt pumps.
7 Good evening, this is the ten o'clock news for April the 15th, 2015. Twenty-two oil wells have been drilled below the Perdido Spar oil platform, and another thirteen wells have been drilled fifteen kilometres away. This brings the total number of wells below the giant spar to thirty-five.
8 This is the World this Weekend on Saturday, October 8th, 2016. More than 46 million barrels of oil have been produced by the Perdido Spar. During the first year of production at the world's deepest spar, an average of 130,000 barrels per day of oil and natural gas were produced by the 35 wells operating from the spar platform.

25 optional listening

26 optional listening

27

If you look at Diagram A, that's the diagram of the oil rig on the left, you can see all the main moving and non-moving parts. The derrick is the actual tower of the oil rig, and stands solidly on the drilling platform. So the derrick and the platform obviously don't move, because they support all the other equipment.

Right, so as Diagram A shows, right at the top of the derrick is the crown block. This is fixed to the top of the derrick and doesn't move up or down, although of course it rotates when the winch pulls or releases the cable. Below the crown block is the travelling block, attached to the cable, which moves up and down and raises or lowers the hook with the drilling equipment attached to it.

The hook at the bottom of the travelling block is attached to a swivel. The top part of the swivel can't rotate, but the lower part can.

The lower part of the swivel is attached to a kelly, as you can see in Diagram B, and the kelly fits into and goes through a turntable. Below the turntable, the kelly is attached to the drill pipe. When the diesel engines are switched on, the turntable rotates, and this makes the kelly rotate. The kelly then makes the drill pipe rotate.

At the bottom of the oil well, a drill collar fits over the drill pipe just above the drill bit. The drill collar helps to weigh down the drill bit. When the drill pipe rotates, this makes the drill bit turn and cut into the rock.

28

[I = Interviewer; AK = Asif Khan]
I: Here I am on the Western Desert oil rig, one of the biggest land rigs in the world, and I'm talking to one of the drilling crew, Asif Khan. Asif, I understand that you were the driller on the day that the first well was drilled from this rig.
AK: Yes, that's right. With my team. We drilled the first well here.
I: So could you talk us through what you and the others did in that first drill.
AK: Well, first of all we made up the drill pipe and the drill bit.
I: Sorry, could you just explain what you mean by 'made up'?
AK: Yeah, what I mean is, we attached the drill pipe to the drill bit, we screwed them together.
I: I see, thanks.
AK: And then we slid the drill collar over the drill pipe so that it sat on top of the drill bit.
I: Right.
AK: And then we made up the pipe with the kelly.
I: Made up? Oh, yes, I remember. You joined the kelly to the drill pipe?
AK: Yes, that's right. So then we lowered the string …
I: Excuse me, what do you mean by 'string'?
AK: The string – you join different parts together to make a long section. So the string is the drill bit, the collar, the drill pipe and the kelly, all joined together. The whole string rotates inside the well hole.
I: Oh right, I see, thanks.
AK: So then we lowered the string – the kelly, the drill pipe, the drill collar and the drill bit – we lowered it through the rotary table until the kelly fitted tightly in the hole in the rotary table.
I: I see.
AK: And then the mud pump was switched on, to make the drilling fluid flow down to the drill bit.
I: Was that to lubricate the drill bit?
AK: That's right. And then the drilling mud hose was checked for leaks.
I: Sorry, I didn't catch that. It's the noise. Could you repeat that, please?
AK: Sure, the mud hose was checked for leaks.
I: Who did that? Did you do it?
AK: No, someone else. The derrick hand did that.
I: Right, and then after the hose was checked …?
AK: Then the rotary table was switched on.
I: You didn't do that?
AK: No, that's the job of the motor hand. He did that.
I: Then you started drilling?
AK: That's right. I slowly tripped in …
I: Sorry to interrupt you, but what does 'tripped in' mean?

AK: Tripped in? That means lowered the drill string into the well. So I tripped in, or lowered the drill bit to the rock below the platform.
I: And that's how the drilling on the first well was begun?
AK: That's right.
I: And how many metres have been drilled so far?
AK: About fifty metres up to now.
I: That's great. Well, Asif, thanks for talking to us.
AK: My pleasure.

Unit 9 Design

29

A: I've just finished testing the new Can-Am Spyder roadster, and comparing it against the Zoomster XL motorcycle.
B: So, what were the main differences?
A: Well, of course, the first thing to mention is the wheels. While the Zoomster has two wheels, the Spyder has three – that's two in front and one behind.
Having three wheels means having three sets of brakes. And here's another obvious and very striking difference between the Spyder and any motorbike, not just the Zoomster. Most motorbikes have a brake pedal for the back wheel, and a lever on the handlebar for the front brake. But the Spyder has a single brake pedal, located on the right side, for all the brakes, and it has no lever. So it was a bit confusing at first, but I soon became accustomed to the new layout.
B: How did the brakes perform in your test?
A: Well, I found that the Spyder's brakes were much more efficient and a great deal more powerful than the Zoomster's.
B: How about speed?
A: Well, I found that the Spyder went just as fast as the Zoomster. The maximum speed of both vehicles was almost the same, at 260 kilometres per hour. So that's the good news.
B: What's the bad news?
A: Well, although the Spyder's speed was as good as the Zoomster's, its acceleration wasn't as good. I would say the Spyder's acceleration was about ten per cent slower.
Another problem is that the two front wheels meant that it was quite a lot harder to turn the handlebars to steer the Spyder.
A further issue was that whenever the Spyder's two front tyres went over a hole or bump in the road, I felt a lot of up-and-down movement in the handlebars. The Zoomster had much better suspension and went over the bumps without too much movement in the bars.
B: So, what about safety?
A: The Spyder is equipped with safety systems which keep all three wheels in contact with the road at all times. So turning round a bend at speed is much more stable and safer than on the Zoomster. On the whole, I felt that the Spyder was much less dangerous than most normal motorbikes. Perhaps that makes it less exciting? I'll leave that for you to decide.
Oh, before I forget, the storage space at the front of the Spyder is very roomy. It's about twice as large as the container on the back of the Zoomster.
B: And what's your overall assessment?
A: All in all, the Spyder proved to be a useful and ultra-safe innovation in the motorcycle market, much safer than the Zoomster. It will perhaps be more attractive to slightly older riders. That's just as well, because the Spyder is about fifty per cent more expensive than the Zoomster.

30 optional listening

31 optional listening

32

... and our design for the new Institute of Maritime Studies has recently been shortlisted as best site in the Learning category at the World Architecture Festival. So, could you all please look at the plan of the site of the new campus? You can see that the complete site is enclosed by water on three sides, and the main highway on the fourth side. Two canals on the east and west sides flow into the large lake on the south side of the site. In other words, there is plenty of water around, which is appropriate for maritime studies.

Right, so I'm now going to point out some of the main buildings of the site. Let's begin with the building in the furthest north-east corner of the site. It's the long, curved building adjacent to a curved and tapering stretch of water. Do you see it? It's right next to the small curved area of water. That's the new administration building.

OK, so I'd like to move on to another building, the hostels where the students will have their accommodation. The student hostels are in the long narrow rectangular building on the opposite side of the curved lake from the admin building.

Right, so the next building that I would like to mention is the Academic Block, which contains the main lecture theatres and classrooms. This is the curved building which looks like a set of teeth. This building is on the opposite side of the road from the student hostels.

Just south west of the Academic Block you can see a rectangular blue pool of water, enclosed in an oval or elliptical building. This is the swimming pool.

Just south of that there are two buildings. One is a doughnut-shaped, or ring-shaped building. This is the Research Centre.

Immediately adjacent to the Research Centre, to the east of it, is a rectangular building. This is the Workshop.

On the opposite side of the sports field, the oval building, pointed at both ends, is the Services Building.

And right next to the lake is a structure which is semi-circular at one end and straight on the other end. This is the campus ship. It's a ship where the maritime students can practise their seafaring skills.

Unit 10 Disasters

33

[P = Pete; J = Jerry; T = Tom; S = Susan; R = Richard; Ja= Jason]
P: Welcome back to Mississippi Calling, I'm Pete Hanson, your host, and we're talking about the terrible event that's dominating all the news channels and all the phone-ins tonight, and that's the tragic collapse of the I-35W bridge over our Mississippi river earlier this evening.
We'd like anyone out there who's a civil engineer, or any kind of technical expert on bridge design or bridge construction to call in, and tell us about your theories, speculations and ideas about why this bridge might have collapsed.
So what might have caused the collapse? We'd like to hear from you. OK, now we have Jerry from Minnesota. What's your idea, Jerry?
J: Well, I think that one or more of the girders might have buckled, Pete ...
P: You're an engineer, Jerry?
J: I'm a technician working in a civil engineering company. I think the collapse might have been caused by a girder buckling.
P: And now we have Tom on the line, from Chicago. What's your take on this, Tom?
T: Well, I reckon the collapse could have been due to metal fatigue.
P: You sound like a bridge engineer too, is that right, Tom?
T: Yeah, I've worked on quite a lot of bridge projects. I think that years and years of the same loads over and over again might have caused some metal fatigue in the truss.
P: And now we have Susan, a civil engineer from Texas. Susan, can you shed some light on this?
S: I'll try, Pete. My own view is that one of the bearings must have corroded and rusted away.

P: You say it must have corroded? You sound pretty certain of that.
S: Yeah, well, I've seen it happen before, on two other bridges that collapsed. When they did the investigation, they concluded that the collapse was caused by corroded bearings.
P: And now we have Richard on the line. Well, Richard, what do you think went on here?
R: Well, I think the collapse could have been caused by thermal shock. It was an extremely hot day and if the bearings and plates were also corroded, the heat could have caused too much contraction.

34

[I = Inspector; CR = Company Representative]
I: Your company should have inspected the bridge annually. Did a competent employee of your company carry out an annual inspection?
CR: The bridge was inspected every year, although it wasn't in 2007.
I: Why not?
CR: The main reason was that there was a lot of construction work going on on the bridge.
I: Well, the bridge should have been inspected in 2007. What did previous inspection reports say?
CR: In 1990 a report stated that there was significant corrosion in its bearings.
I: Were the bearings repaired or replaced immediately afterwards?
CR: No, I'm afraid nothing happened as a result of the report.
I: Well, the bearings shouldn't have been left on that bridge. If your company had replaced the bearings, maybe the bridge wouldn't have collapsed. What did other reports on the bridge say?
CR: A 2001 inspection stated that there was cracking in some of the girders.
I: Was any action taken after that report? The cracks should have been drilled to stop them from spreading.
CR: Yes, this remedial action was carried out.
I: But that's not enough. Support struts should also have been added to the cracked girders to prevent any more cracking.
CR: Yes, this was done after the report.
I: All right. What about 2006? There was an inspection then. What did that find?
CR: Signs of metal fatigue were observed in the bridge. The report mentioned the metal fatigue.
I: In my opinion, the bridge should have been closed then, in 2006, immediately after the signs of metal fatigue were discovered. But of course that never happened.
CR: No, that's correct.
I: The bridge should have been replaced immediately.
CR: Yes, you're right.
I: Did anything happen after that inspection in 2006?
CR: Yes, we planned to carry out some steel reinforcement on the bridge. But the project was cancelled.
I: Why?
CR: We found that the reinforcement work might have weakened the bridge.
I: You shouldn't have cancelled the steel reinforcement. You should have found a way to do it without weakening the bridge.
CR: Yes.
I: Look, in our investigation, we've found a design error in the gusset plates which connect the girders together in the truss structure. The plates were too thin to support the girders. You should have discovered this error in one of your annual inspections. Did you?
CR: No, we didn't.
I: And the undersized gusset plates should have been replaced with larger ones. Were they?
CR: No, they weren't.
I: Well, if you had replaced the gusset plates, the bridge would probably not have collapsed.

Unit 11 Materials

35

[RO = Ramón Ortega; AW = Albert Weston]
RO: Ramón Ortega.
AW: Hello, good morning, Ramón. This is Albert Weston. How are you?
RO: Ah yes, Albert. How are you doing? Thanks very much for coming along yesterday. The team were very impressed with your presentation.
AW: Oh good, I'm glad to hear it. It was good to meet the team.
RO: Yes, and they really liked the look of the new football boot that you've designed. We all think that maybe we'll buy your boots for the next season. But we'll have to look into it a little bit more, discuss the price, and so on.
AW: Great. That's why I'm phoning, in fact. I was wondering what the next move is. Would you like me to send you a formal proposal?
RO: Yes, that would be excellent. In the proposal, just summarise the main points you made in your presentation yesterday. Give a bit of technical background about the properties of the materials used in the boot. And of course confirm your unit price, delivery dates and so on.
AW: OK, will do. I'll get the proposal to you by special delivery tomorrow morning.
RO: Very good. Thanks, Albert. We'll see you soon, I hope.
AW: Of course. I look forward to it. Bye now.
RO: Bye.

36

Thank you for clicking on the link to find out more about the materials we use in the revolutionary new DesignerSport football boot.

The upper or top part of the boot is made of a combination of two materials, carbon fibre and aramid fibre. Carbon fibre of course is a very light and flexible material, able to bend easily in all directions. And that's great for comfort and ease of movement. But the player also needs some protection against impact. I'm sure you'll remember the metatarsal injuries that David Beckham and Wayne Rooney suffered in their feet. So that's why we have added aramid fibre to the upper part of the shoe. Aramid fibre is strong in tension, which means that it doesn't stretch when another foot smashes into it. This makes the upper impact resistant. The player's foot is completely protected from injury from outside the boot.

Inside the boot we've put a generous amount of padding, made of polyurethane foam. This material is very soft, but also highly impact absorbent, which means that if the boot strikes (or is struck by) something hard, the padding absorbs the blow and reduces its impact on the foot.

Finally, we have the sole plate on the bottom of the boot. This is made of thermoplastic polyurethane, or TPU, which is a tough plastic, which means that it can't be broken or split by pressure or impact of any kind. But it's also very elastic, which means that it can bend or twist out of shape, and then return to its original shape immediately.

37 optional listening

38 optional listening

39

A: OK guys, I've called this meeting to discuss our plans for our national team for the next-but-one Olympics. We need to make some decisions soon on two important issues – and I'm sure you know what I'm talking about. So let's start with the first issue – equipment. What about running shoes for the 100, 200 and 400 metres?

B: Well, I think we need a more lightweight model. Last year's model gives good impact resistance in the sole, but it's not light enough compared with what other teams have.

C: Why don't we try the new Flite shoes? I've tried them out and they withstand impact extremely well. But they're also incredibly lightweight. They weigh 67 grams each, 40 percent less than last year's model.

A: OK, that sounds good, Jane. We'll look into that. Right, another equipment issue is the long-distance shoes for the marathon. Any ideas?

B: Yes, we need to find a shoe than can tolerate very wet roads and resist slipping. Last year's model doesn't have enough grip on wet surfaces.

D: Let's try using the newest Marathonites. They're made by the same Japanese company that designed shoes for the rain-soaked marathon of the Athens Games. They have good impact absorbency, but the most important property is that they're totally slip resistant on wet surfaces.

A: Thanks for that suggestion, Anil. Can you look into these shoes a bit more and do me a report? Thanks. All right, I've just one more equipment issue before we move on. What about swimsuits for our swimmers? Are they aerodynamic enough?

E: You mean hydrodynamic, don't you?

A: Yes, that's right, they're racing through water, not air. So, can we improve the hydrodynamic properties of our swimsuits?

B: We need to find a material that reduces drag in the water. Perhaps a material like a shark's skin?

C: Hmm, shark's skin. Can I make a suggestion? We could look at the new SpeedShark swimsuit. The manufacturers claim that it is 10 percent more hydrodynamic than other models.

A: Good, why don't we look into that? But we'll have to be careful here. The Olympic Committee may put a ban on new materials for swimsuits, so we'll keep an eye on that, OK? Right, so let's move on to our second important issue of the meeting, namely training. Are there any suggestions for using new technology to improve team training?

B: I would suggest that we need to invest more in sensors that are able to tell us how well each athlete is performing.

D: How about starting with the rowing team? There are very good sensors now that you attach to the rowing blades. They're capable of sending accurate information in real time to the coach.

A: Can they measure the force that the rowers use in each stroke?

D: Yes, they have the capability of providing data on both force and speed.

A: Excellent. Any other suggestions? What about for our sailing team?

B: We need a device that's capable of reading wind speed and wind direction, and presenting the information clearly to the sailor.

E: In the Beijing Games they used a Doppler lidar system. It scans the sea with laser beams. It has the ability to provide a real-time readout of wind speed and direction.

A: Very good. Let's look into all these suggestions and make a full report.

Unit 12 Opportunities

40

[S1 = Scientist 1; S2 = Scientist 2; V = Voice Message]

S1: Hey, something's coming through. Someone's speaking. It's a V from the future. Someone's speaking to us from the year 2060. Can you believe that? Shhh. Let's listen.

V: ... the world's temperature has risen by eight degrees Celsius since your time ... fires have burnt down huge areas of forest ... most of the world's forests have now been destroyed ... the Arctic ice cap has completely disappeared ... the glaciers on the world's mountains have melted and turned into rivers ...

S2: What was that? Did you catch it? What did it say?

S1: Something about mountain glaciers. He says that they've all melted.

V: ... sea levels around the world have risen by one point two metres since your time ... many low-lying countries have been flooded ... tropical cyclones have destroyed large parts of many of the world's major cities ... water in many villages has dried up and their populations have died because of the drought ... every year the emissions of carbon dioxide into the atmosphere have increased ...

S2: Did you hear what he said just then?

S1: Yeah, he's talking about carbon dioxide now. Emissions have increased year by year.

V: ... now emissions of carbon dioxide have risen to eighty gigatonnes per year ... and the concentration of carbon dioxide in the atmosphere has also gone up every year since your time ... now the concentration of carbon dioxide has risen to two thousand parts per million ...

41

V: Your society should have reduced your ... of oil and other fossil fuels. You should have invested more in renewable energy. Your governments shouldn't have encouraged cheap air flights; instead, they should have put higher taxes on air fuel to ... the cost of air travel. Everyone should have ... their own energy in their homes. They should have ...[wind turbines and solar panels on their houses. Why didn't your society and governments do these things? If you had ...out these actions, the world's temperature probably would not have ... by eight degrees Celsius. If your government had ... better decisions, the sea level would probably not have ... by one point two metres, and low-lying areas would have been .... .

42

[B = Boss S = Scientist]

B: You heard the message from the future, didn't you?

S: Yes, that's right.

B: So could you please summarise very briefly what will have happened by the year 2060, according to the message?

S: Certainly. By 2060, CO_2 emissions will have risen to 80 gigatonnes per year. CO_2 concentrations in the atmosphere will have increased to 2000 ppm. The world's temperature will have gone up by 8 degrees from today's levels. As for the sea level ...

Pearson Education Limited

Edinburgh Gate
Harlow
Essex CM20 2JE
England

and Associated Companies throughout the world.

www.pearsonlongman.com

First published 2011

ISBN: 978-1-4082-2947-7

Set in Adobe Type Library Fonts

Printed in Slovakia by Neografia

Acknowledgements

The publishers and author would like to thank the following people and institutions for their feedback and comments during the development of the material:

Francis McNeice, Morocco; Simon Turner, France; Gundula Wilke, Germany; Peter Amoss, Germany; Maruisz Starak, Poland; Iwona Galazka, Poland; Ahmed Motala, UAE.

The author would like to thank Avtar Bonamy for her inexhaustible artesian aquifer of clear spiritual spring water. Her support, encouragement and wisdom supplied the Avtur that maintained *Technical English 3* at a safe altitude. The author is also profoundly grateful to the excellent Tony Garside, whose clear perceptions and sound judgment provided scaffolding, earthing and satellite navigation until the project was safely berthed.

The publisher would like to thank the following for their kind permission to reproduce their photographs:

(Key: b-bottom; c-centre; l-left; r-right; t-top)

ADNEC (Abu Dhabi National Exhibitions Company): 70tl; **Alamy Images:** Andrew Twort 81 (C), Arcticphoto 82bl, Davo Blair 48, IML Image Group Ltd 18tr, Inspirestock Inc 86 (1), Jim Parkin 62, Michael Willis 98c, Mike Hill 90-91, Peter Huggins 81 (D); **Alan Holden:** 71l; **alveyandtowers.com:** 12 (bumper), 36, 52 (x-ray), 86 (3); **Art Directors and TRIP photo Library:** Helene Rogers 14l, 81 (A), 81 (B); **Atkins:** 81 (4); **© Belkin International:** 81 (2); **Camera Press Ltd:** LAYER WERNER / JACANA 82br, PIEL PATRICK / GAMMA / Eyedea Presse 82bc; **Christopher Charles Benninger Architects Private Limited:** 72 (1), 72 (2), 72r; **Corbis:** Justin Lane / epa 18tl, Michel Setboun 4; **Courtesy of Nokia:** 80; **CROWN Gabelstapler GmbH & Co. KG:** 99cr; **David Kimber:** 68; **DK Images:** JR Marshall 42 (turbine); **Dominic Schindler Creations GmbH:** 99tr; **Ecotricity Group Ltd:** 92; **Fiberline composites A/S / Paul O. Elmström :** 10bl; **Frog Design Inc:** 69; **Gaia & Gino:** Serdar Samli 81 (3); **Getty Images:** Chris Ratcliffe / Bloomberg 111, David Lees 37, Eightfish 86 (2), Ira Block 86 (4), Robert Nickelsberg 56, SNSM / AFP 82tr, STF / AFP 35t, 35b, Zac Macaulay Photography Ltd, 26tl; **Ronald Grant Archive:** Paramount Pictures 33; **iStockphoto:** 12 (drainpipe), 71r, 81 (1), 99cl, 99bl; **Makaoto Funamizu, http://petitinvention.wordpress.com:** 98r; **Pearson Education Ltd:** Photodisc / StockTrek 98bl; **Photolibrary.com:** Creatas 52tr, Imagesource 40, 84, Nick White & Fiona Jackson-Downes 30, Photodisc 42 (rig), Thomas Dressler 42 (panel), Volket Steger 26tr; **Press Association Images:** 4tr, 77, AP Photo / Charles Dharapak 74b, AP Photo / John Weeks III 74t, Neal Simpson 82tl; **Rex Features:** 24, 32, 82tc, 88, 98tl, GM 10tl, KPA / Zuma 96, Pekka Sakki 12 (bag), S. Corvaja 20; **Scansis AS, Scan-ICE®:** 10r; **Science Photo Library Ltd:** Martin Bond 34; **Slipform International Ltd:** 60b; **Statoil:** 60t; **STILL Pictures The Whole Earth Photo Library:** Biosphoto / Duret Agnès 42 (dam); **SuperStock:** age fotostock 70c, RelaXimages 70r; **Thinkstock:** Comstock 12 (switch), Digital Vision 52tl, George Doyle & Ciaran Griffin / Stockbyte 97, Hemera Technologies 87, 99tl, iStockphoto 12 (bottle), 12 (telephone), 14tr, liquidlibrary 14br; **TopFoto:** 52 (cameras); **www.ziplux.com:** Walen Souza Cruz Jr 99br

Cover images: *Front:* **Alamy Images:** Technology And Industry Concepts

All other images © Pearson Education

Every effort has been made to trace the copyright holders and we apologise in advance for any unintentional omissions. We would be pleased to insert the appropriate acknowledgement in any subsequent edition of this publication.

Picture Research by Kevin Brown

Designed by Keith Shaw